Game-Theoretic Learning and Distributed Optimization in Memoryless Multi-Agent Systems

Tatiana Tatarenko

Game-Theoretic Learning and Distributed Optimization in Memoryless Multi-Agent Systems

Springer

Tatiana Tatarenko
TU Darmstadt
Darmstadt, Germany

ISBN 978-3-319-88039-6 ISBN 978-3-319-65479-9 (eBook)
DOI 10.1007/978-3-319-65479-9

Softcover reprint of the hardcover 1st edition 2017

Printed on acid-free paper

This Springer imprint is published by Springer Nature
The registered company is Springer International Publishing AG
The registered company address is: Gewerbestrasse 11, 6330 Cham, Switzerland

Abstract

Learning in potential games and consensus-based distributed optimization represent the main focus of the work. The analysis of potential games is motivated by the game-theoretic design, which renders an optimization problem in a multi-agent system a problem of potential function maximization in a modeled potential game. The interest to distributed consensus-based optimization is supported by growing popularity of dealing with networked systems in different engineering applications.

This book investigates the algorithms that enable agents in a system converging to some optimal state. These algorithms can be classified according to information structures of systems. A common feature of the procedures under consideration is that they do not require agents to have memory to follow the prescribed rules. A general learning dynamics applicable to memoryless systems with discrete states and oracle-based information is presented. Some settings guaranteeing an efficient behavior of this algorithm are provided. A special type of such efficient general learning procedure, called logit dynamics, is considered. Further, the asynchronous and synchronous logit dynamics is extended to the case of games with continuous actions. Convergence guarantees are discussed for this continuous state dynamics as well. Moreover, the communication- and payoff-based algorithms are developed. They are proven to learn local optima in systems modeled by continuous action potential games. The stochastic approximation technique used to investigate the convergence properties of the latter procedures is also applied to distributed consensus-based optimization in networked systems. In this case, the stochastic nature of the proposed push-sum algorithm allows a system to escape suboptimal critical points and converge almost surely to a local minimum of the objective function, which is not assumed to be convex.

Contents

Chapter 1
Introduction

1.1 Motivation of Research

Due to the emergence of distributed networked systems, problems of cooperative control in multi-agent systems have gained a lot of attention over the recent years. Some examples of networked multi-agent systems are smart grids, social networks, autonomous vehicle teams, processors in machine learning scenarios, etc. Usually in such system there is a global objective to be achieved by appropriate local actions of agents. Such optimization problems can be solved centrally. However, for a centralized solution a central controller (central computing unit) is required, to collect the whole information about the system and to solve the optimization problem under consideration. This approach has limitations. Firstly, systems with such settings are sensitive to the failure of the central unit. Secondly, the information exchange is costly, since agents need to transmit their local information to the central unit and to receive the instructions from it. Moreover, due to a large network's dimension, the optimization problem can become computationally infeasible for the central controller. Finally, it may be no resources to incorporate a central computing unit into the system. Thus, in networked multi-agent systems agents are motivated to optimize a global objective without any centralized computation by taking only the local information into account. Various approaches have been proposed so far to deal with this problem, including reinforcement learning, game-theoretic modeling, and distributed optimization algorithms.

This work focuses mainly on the *game-theoretic approach* to optimization in multi-agent systems. One of the reasons to use this approach in different applications is that one can design games, where a subset of Nash equilibrium states corresponds to the system optimal states that are desirable from a global perspective. In this context *game-theoretic learning* is a topic of specific interest. It concerns the analysis of the distributed adaptation rules that should be based on the local information available to agents and provide convergence of the collective behavior to an optimal solution of the global problem. On the other hand, there

T. Tatarenko, *Game-Theoretic Learning and Distributed Optimization in Memoryless Multi-Agent Systems*, DOI 10.1007/978-3-319-65479-9_1

are multi-agent optimization problems falling naturally under this game-theoretic framework. One can think of a problem in a networked system (for example, cooperative routing), where the cost of each agent depends not only on her own action (distribution of demand to be transmitted over the network), but also on the joint actions (decisions) of other agents (congestion on chosen routes). This coupling evokes a game-theoretic formulation, since the local interests depend on the global behavior. We assume that the objective in the system is to minimize the overall cost of the agents (the social cost of routing), whereas each agent is only aware of her own cost. Such an optimization problem can be considered a special case of a *distributed optimization* problem, where agents need to come to a consensus over their estimations of the optimal state. Moreover, this consensus should meet the global optimizer of the system. Thus, an average dynamics needs to be found to allow agents utilizing the local information in such an efficient way that would guarantee dynamics minimizing the overall cost.

As we can see, the local information plays an important role in both game-theoretic learning and in distributed optimization algorithms. This work provides an appropriate dynamics for searching an optimum in multi-agent systems given different information settings in these systems. Moreover, to restrict the system resources required to perform an algorithm, the present work focuses on *memoryless procedures*, in which agents base their decisions (their action choices) only on the current information and do not need to store their inputs and outputs on the previous steps.

The book begins with the introduction of some important concepts in the game theory, comparison of possible information structures in systems, and overview of the existing relevant literature. All these issues are provided in Chap. 2. In this chapter a general paradigm and motivation of the game-theoretic approach and the distributed optimization in multi-agent systems are highlighted.

Motivated by the discussion in Chap. 2, the work proceeds to considering memoryless multi-agent systems designed by means of potential games. Chapter 3 studies systems with the so-called oracle information, where each agent can calculate her current output given any action from her action set. Firstly, the chapter deals with discrete action potential games. A general memoryless stochastic procedure with specific properties is proposed to guarantee convergence in total variation of joint actions to a potential function maximizer, which under an appropriate game design coincides with a global optimizer in the centralized problem. The convergence rate is estimated for this procedure as well. Moreover, to accelerate this rate, this work provides some important insights into the finite time behavior of the procedure. The analysis is based on the theory of *Markov chains with finite states*. Background material on this topic is discussed in Appendix A.2. As special cases of such general memoryless oracle-based algorithm, the logit dynamics and its synchronous version are considered. Some optimal settings for the parameters of these dynamics are established, under which the learning processes demonstrate an efficient performance. Moreover, the difference between the asynchronous and synchronous oracle-based learning dynamics is explained in terms of required information structures. However, the assumption of discrete action sets does not hold in many applications of multi-agent systems. For instance, in the case of energy-efficient power control and power allocation in large wireless networks users can transmit at different power levels and their payoffs are continuous functions of their chosen transmit power profiles. Analogous situation is encountered in "smart grid" setting, that is currently gaining a lot of attention in power generation and management of electricity markets. Indeed, agents in such systems deal with continuous variables by making decisions on how much power needs to be generated, consumed, or stored during the day. That leads again to a game-theoretic design with continuous action sets. This motivates the second part of Chap. 3 that is focused on the logit dynamics extended to continuous actions. Although stochastic stable states in the standard asynchronous logit dynamics can be studied by means of the detailed balance equation, the analysis of its synchronous continuous version requires more sophisticated techniques. Chapter 3 introduces a new approach for characterizing the stationary distribution of the synchronous logit dynamics to demonstrate stochastic stability of system optima in this procedure. This approach uses the results on *continuous action Markov chains*, which are summarized in Appendix A.3. The analysis of finite time behavior of the continuous dynamics is presented. This enables setting up such algorithms' parameters that are justified by the algorithms' simulations demonstrating a fast approach to an optimal state.

Multi-agent systems with communication- and payoff-based information structures are considered in Chap. 4. The chapter begins by introducing some preliminaries on *stochastic approximation theory* that is used further for analysis of those algorithms. First, the communication-based push-sum protocol, which has been recently applied to convex distributed optimization, is adapted to the case of *non-convex* functions, the sum of which is to be minimized in a distributed manner. The advantage of this memoryless protocol in comparison with other protocols is that it does not require double stochastic communication matrices to guarantee convergence of agents' local arguments to a consensus point. This makes potential applications of the procedure much broader. The convergence of the initial deterministic procedure to a critical point of the objective function[1] is proven in non-convex settings. Moreover, it is demonstrated that a stochastic term added to the deterministic procedure allows the procedure to escape critical points different from local minima and, thus, to converge to a local optimum almost surely. The convergence rate of the stochastic algorithm is also provided. This convergence result motivates the application of this protocol to communication-based learning in potential games. Chapter 4 presents the version of the push-sum protocol that can be efficiently applied to a multi-agent system, in which an optimization problem is modeled by means of a potential game. Additionally, the chapter considers the payoff-based learning algorithm as well. In this case agents do not have access either to the structural properties of their payoffs or to the communication channels between them. The only information available is the currently chosen local actions

[1]Recall that *critical points* of a smooth function are zeros of its gradient.

and corresponding individual payoffs. It turns out that one can make use of the stochastic approximation technique, similar to one used in the analysis of the push-sum protocol, to develop a payoff-based learning that converges to a local minimum of the potential function almost surely. Moreover, a similar approach can be applied to completely distributed and uncoupled learning of Nash equilibria in potential games with concave utility functions. As an example of such games, a game between users of electricity markets is considered, where users' goal is to converge to a stable joint strategy over time given only some local information about the actual cost values.

Note that the book is divided into the chapters according to the mathematical tools required for the corresponding investigations. Thus, the analysis in Chap. 3 is based on the general theory of Markov chains with discrete and continuous states. This theory is applied to study the behavior of oracle-based memoryless learning procedures in potential games. The algorithms presented in Chap. 4, both for game-theoretic learning and distributed optimization, are related to each other by stochastic approximation techniques that were used to study them. The basic definitions of convergence types in probabilistic spaces relevant to the theoretic part of the book are presented in Appendix A.1.

1.2 List of Notations

$[N]$	The set $\{1, \dots, N\}$
$\{A_i\}_i$	The set of the elements A_i indexed by i
$\lvert\mathcal{R}\rvert$	The cardinality of the set $\mathcal{R}$
$2^{\mathcal{R}}$	The set of all subsets of the set $\mathcal{R}$
$\mathbb{R}$	The set of real numbers
$\mathbb{R}^+$	The set of nonnegative real numbers
$\mathbb{Z}$	The set of integers
$\mathbb{Z}^+$	The set of nonnegative integers
P, p_{ij}	The matrix and its elements, respectively
$\boldsymbol{a}$	The vector value
a^i	The ith coordinate of the vector $\boldsymbol{a}$
$(\boldsymbol{a}_1, \boldsymbol{a}_2)$	The dot product of two vectors $\boldsymbol{a}_1$ and $\boldsymbol{a}_2$
$\lVert\boldsymbol{a}\rVert$	The Euclidean norm of the vector $\boldsymbol{a} \in \mathbb{R}^d$, namely $\lVert\boldsymbol{a}\rVert = \sqrt{\sum_{i=1}^d a_i^2}$
$\lVert\boldsymbol{a}\rVert_{l_1}$	The l_1 norm of the vector $\boldsymbol{a} \in \mathbb{R}^d$, namely $\lVert\boldsymbol{a}\rVert = \sum_{i=1}^d \lvert a^i\rvert$
$\sigma(X)$	The Borel sigma-algebra of the space X
$P(\cdot, \cdot)$	The stochastic kernel defined on the Borel space $(X, \sigma(X))$, namely the function $P(\cdot, \cdot) : X \times \sigma(X) \to [0, 1]$

$\|P(\cdot,\cdot)\|_{TV}$	The norm of the stochastic kernel $P(\cdot,\cdot)$ defined on the Borel space $(X,\sigma(X))$ by total variation topology, namely $\|P\|_{TV} = \sup_{x\in X}\int_X \|P(x,dy)\|$
$\delta_a(x)$	The Dirac delta function, namely $\delta_a(x) = \begin{cases} +\infty, & x = a \\ 0, & x \neq a. \end{cases}$
ρ	The metric of a metric space X:$\rho(x,B) = \inf_{y\in B}\rho(x,y)$, $x \in X, B \subset X$
$f^{(k)}$	Derivatives of the kth order of the function f
$f,g:\mathbb{R}^m \to \mathbb{R} : f(x) = O(g(x))$ as $x \to a$	Big O notation: $\lim_{x\to a}\frac{\|f(x)\|}{\|g(x)\|} \leq K$ for some positive K
$f(x) = \Omega(g(x))$ as $x \to a$	Big Ω notation: $f(x) \geq Kg(x)$ for some positive K and x sufficiently close to a
$f(x) = \Theta(g(x))$ as $x \to a$	Big Θ notation: $K_1 g(x) \leq f(x) \leq K_2 g(x)$ for some positive K_1, K_2 and x sufficiently close to a
$\bigotimes$	Kronecker product of matrices
$\mathrm{cl}(A)$	Closure of the set A
I_N	The N-dimensional identity matrix
$\mathbb{1}_N$	The N-dimensional vector of unit entries

Chapter 2
Game Theory and Multi-Agent Optimization

In this section, a background of the game theory with applications to optimization in multi-agent systems is presented. Game theory is "the study of mathematical models of conflict and cooperation between intelligent rational decision-makers" [Mye91]. This study was primarily developed as a useful tool for the analysis in mathematical economics [NM44]. Originally, it addressed two-player zero-sum games, where any gain of one agent is equal to a loss for another one. Today, however, game theory is proven to provide a valuable technique to study and control various complex systems where behavioral relations between agents play an important role in the efficiency of systems' evolution [Sie06].

Since the game theory deals with interacting decision-makers, the motivation to use a game-theoretic approach in analysis and characterization of multi-agent systems in engineering applications is quite clear. The game-theoretic approach in control theory enables one to overcome the analytical difficulties of dealing with overlapping and partial information in classical design of local control laws [CNGL08, LM13, TA84]. Many control problems, such as consensus, network formation and routing, target assignment, coverage problems, optimal power transmission, or resource allocation problems, arise in engineering systems. To all of them a game-theoretic approach can be efficiently applied [AMS07, Int80, SFS09, SZPB12, SBP06].

2.1 Game Theory

2.1.1 Introduction to Game Theory

We start by considering a multi-agent system consisting of N agents, $[N] = \{1, \ldots, N\}$. Each agent $i \in [N]$ has an action set A_i, which can be either finite (discrete) or infinite (continuous). The set of joint actions is denoted by $\boldsymbol{A} =$

T. Tatarenko, *Game-Theoretic Learning and Distributed Optimization in Memoryless Multi-Agent Systems*, DOI 10.1007/978-3-319-65479-9_2

$A_1 \times \cdots \times A_N$. We say that there is a game between the agents, if each agent i has a utility function $U_i : \boldsymbol{A} \to \mathbb{R}$, whose value depends not only on the action $a^i \in A_i$ of the agent i, but also on the joint action a^{-i} of all other agents, namely on $a^{-i} = (a^1, \ldots, a^{i-1}, a^{i+1}, \ldots, a^N)$, i.e.,

$$U_i(\boldsymbol{a}) = U_i(a^i, a^{-i}) = U_i(a^1, \ldots, a^N).$$

In this case the agents are considered the players in the game $\Gamma = (N, \{A_i\}_i, \{U_i\}_i)$.

Example 2.1.1 An example of a discrete action game with two players is the well-known prisoner's dilemma. The possible actions of the players and their utilities can be presented by the table in Fig. 2.1. The following interpretation is standard. Two players are two criminals who are arrested. The following bargain is proposed to the prisoners, who have no way to communicate with each other. Each prisoner can either betray (defect) the other or cooperate with him and remain silent. The corresponding punishment in four different cases of prisoners' behavior is described by Fig. 2.1. If both prisoners cooperate, each of them will serve 1 year in prison, if they both defect, each of them will serve 2 years in prison. If one of the prisoners defects (D), whereas another cooperates (C), that one, who defects, will go free, but that one, who cooperates, will serve 3 years in prison. Thus, the outcome for each prisoner depends not only on his action, but also on the action of his opponent. In other words, there exists the game $\Gamma = (2, \{A_i\}, \{U_i\})$ between the prisoners, where $A_i = (C, D)$ for $i = 1, 2$, and utility functions are equal minus the number of years in the corresponding punishment, i.e.,

$$U_i(a^i, a^{-i}) = \begin{cases} -1, \text{ if } a^i = C, a^{-i} = C, \\ 0, \text{ if } a^i = D, a^{-i} = C, \\ -3, \text{ if } a^i = C, a^{-i} = D, \\ -2, \text{ if } a^i = D, a^{-i} = D. \end{cases}$$

Example 2.1.2 An example of a continuous action game can be found in a smart grid system consisting of N electric vehicles [AMS07]. Let us assume each vehicle in the smart grid to have an energy demand $D_i > 0$ as well as a battery capacity

Prisoner 2 / Prisoner 1	cooperate	defect
cooperate	*Each serves 1 year*	*Prisoner 1: 3 years Prisoner 2: goes free*
defect	*Prisoner 1: goes free Prisoner 2: 3 years*	*Each serves 2 year*

Fig. 2.1 The prisoner's dilemma

$C_i > 0$, $i \in [N]$ ($D_i \leq C_i$). The total charging period of all N vehicles is divided into a fixed number of T discrete intervals (e.g., 1-h intervals). Each vehicle needs to formulate a charging plan $\mathbf{x}_i = [x_i^1, \ldots, x_i^T]$, where x_i^t, $t \in [T]$, is the ith vehicle's requested charging rate for the tth time interval. Each vector $\mathbf{x}_i$, $i \in [N]$, must fulfill the following constraint:

$$D_i \leq \sum_{t=1}^{T} x_i^t \leq C_i \tag{2.1}$$

in order to satisfy the demand of the ith vehicle, but not to exceed the capacity of its battery. At any time slot $t \in [T]$ each vehicle $i \in [N]$ receives only the fraction ρ_t of its requested rate x_i^t. This fraction depends on the overall rate requested at the step t, namely

$$\rho_t = f\left(\sum_{k=1}^{N} x_k^t\right).$$

Thus, the total energy received by the vehicle i over the whole time period is

$$U_i(\mathbf{x}_i, \mathbf{x}_{-i}) = \sum_{t=1}^{T} x_i^t \rho_t = \sum_{t=1}^{T} x_i^t f\left(\sum_{k=1}^{N} x_k^t\right).$$

Since the utility U_i above depends not only on the charging plan $\mathbf{x}_i$ of the vehicle i, but also on the plans of all other vehicles $\mathbf{x}_{-i}$, one concludes that there exists the game $\Gamma = (N, \{\mathbf{X}_i\}, \{U_i\})$ between vehicles, where the action set $\mathbf{X}_i$ of the ith player (the set of feasible plans of the ith vehicle) is the set of the vectors satisfying (2.1).

The examples above deal with *pure strategy* games. It means that players choose only one action from their action sets. In other words, each player chooses some action with probability one. If players are allowed to make a decision according to some *independent* probability distributions, then it said that players "mix" their strategies and one faces a *mixed strategy game*. Formally, some game $\tilde{\Gamma} = (N, \{\Sigma_i(A_i)\}, \{\tilde{U}_i\})$ is a mixed strategy game based on a pure strategy game $\Gamma = (N, \{A_i\}, \{U_i\})$ (also called *extension* of Γ to mixed strategies), if $\Sigma_i(A_i)$ is the set of probability distributions defined on the set A_i and $\tilde{U}_i$ is the mathematical expectation of U_i under the joint distribution σ from the set $\Sigma(\mathbf{A}) = \Sigma_1(A_1) \times \cdots \times \Sigma_N(A_N)$. More specifically, given a discrete action pure strategy game Γ, the ith player's utility function in the corresponding mixed strategy game $\tilde{\Gamma}$ is

$$\tilde{U}_i(\sigma) = \tilde{U}_i(\sigma^1, \ldots, \sigma^N) = \sum_{\boldsymbol{a} \in \boldsymbol{A}} U_i(\boldsymbol{a})\sigma^1(a^1) \cdot \ldots \cdot \sigma^N(a^N),$$

where $\sigma^j \in \Sigma_j(A_j)$ is a mixed strategy of the jth player (some fixed probability distribution over A_j, namely $\sigma^j(a^j) \in [0, 1]$ for any $a^j \in A_j$, and $\sum_{a^j \in A_j} \sigma^j(a^j) = 1$). If some continuous action game in pure strategies is considered, then the utility functions $\{\tilde{U}_i\}$ in its mixed strategy version are expressed by the Lebesgue integrals, namely

$$\tilde{U}_i(\sigma) = \int_A U_i(\mathbf{x})d\sigma(\mathbf{x}),$$

where $\sigma(\mathbf{x})$ is the players' joint distribution of the mixed strategies.

2.1.2 Nash Equilibrium

Assuming that players in a pure (mixed) strategy game are selfish and make decisions independently of each other, the stable outcome of the game would be one in which each player chooses the best response to all other players' strategies.

Definition 2.1.1 In a pure (mixed) strategy game $\Gamma = (N, \{A_i\}_i, \{U_i\}_i)$ $(\tilde{\Gamma} = (N, \{\sigma_i(A_i)\}, \{\tilde{U}_i\}))$ an action $a^{*i} \in A_i$ (mixed strategy $\sigma^{*i} \in \sigma_i(A_i)$) is a *best response* for the player $i \in [N]$ to the joint action of other players a^{-i} (joint mixed strategy of other players σ^{-i}), if

$$U_i(a^{*i}, a^{-i}) = \max_{a^i \in A_i} U_i(a^i, a^{-i}) \qquad (\tilde{U}_i(\sigma^{*i}, \sigma^{-i}) = \max_{\sigma^i \in \sigma_i(A_i)} \tilde{U}_i(\sigma^i, \sigma^{-i})).$$

A Nash equilibrium is a joint action set (joint mixed strategy) in which all players choose their best responses.

Definition 2.1.2 A joint action $\boldsymbol{a}^* \in \boldsymbol{A}$ (joint mixed strategy $\boldsymbol{\sigma}^* \in \sigma(\boldsymbol{A})$) is a *Nash equilibrium* (*Nash equilibrium in mixed strategies*), if the action a^{*i} (the mixed strategy σ^{*i}) is a best response to the joint action of other players a^{*-i} (joint mixed strategy of other players σ^{*-i}) for each player $i \in [N]$.

The crucial question to be investigated in any game is the existence of a stable state, namely of a Nash equilibrium. For example, in the game formulated for the prisoner's dilemma in Example 2.1.1 the Nash equilibrium exists. It is the joint action (D, D). There is no incentives, neither for Prisoner 1 nor for Prisoner 2, to deviate from the strategy "defect" given that the opponent defects as well. Indeed, if one of them cooperates in this case, then he gets 3 years of the prison (see Fig. 2.1). Note also that the joint action (D, D) is the unique Nash equilibrium in pure strategies, since for all other joint actions there exists also a possibility for one of the prisoners to improve his outcome by changing the action.

There are examples of discrete action games that possess no Nash equilibrium in pure strategies. Let us consider a game of two players defined by the following matrix:

	B_1	B_2
A_1	$(-1, 1)$	$(1, -1)$
A_2	$(1, -1)$	$(-1, 1)$

As before the rows correspond to the actions of the first player (A_1 or A_2), whereas the columns define the actions of the second player (B_1 or B_2). On the intersection stay the values of the corresponding utility functions (U_1, U_2). It is straightforward to check that the game above does not have a Nash equilibrium in pure strategies. However, the joint mixed strategy of the players $\boldsymbol{\sigma} = (\sigma^1, \sigma^2)$, in which each of them chooses one of his actions with the same probability $1/2$, i.e.,

$$\sigma_1 : \qquad a^1 = \begin{cases} A_1 \text{ with probability } 1/2, \\ A_2 \text{ with probability } 1/2 \end{cases}$$

and

$$\sigma_2 : \qquad a^2 = \begin{cases} B_1 \text{ with probability } 1/2, \\ B_2 \text{ with probability } 1/2, \end{cases}$$

is the mixed strategy Nash equilibrium in the mixed strategy game defined on all possible probability distributions over the sets $\{A_1, A_2\}$ and $\{B_1, B_2\}$. Moreover, *any finite discrete action game contains at least one Nash equilibrium in mixed strategies* [Nas50].

For games with continuous actions there exist other methods to investigate the questions of existence and uniqueness of Nash equilibrium. For instance, in the game between vehicles in the smart grid from Example 2.1.2, one can use the result presented in the work [Ros65] to demonstrate that the Nash equilibrium exists, if the function U_i is concave in $\mathbf{x}_i$ for each fixed $\mathbf{x}_{-i}$ for all $i \in [N]$.

2.1.3 Potential Games

In this subsection an important class of games called *potential games* is described. This class of games is introduced in the seminal work [MS96b]. Potential games have recently gained a lot of popularity in various technical applications of the game theory. One of the most attractive properties of any potential game with finite number of agents is that such game always possesses a Nash equilibrium, given the existence of a maximum of the potential function on the set of joint actions.

Let us consider a game $\Gamma = (N, \{A_i\}, \{U_i\})$ with $N < \infty$ players. Each player i has an action set A_i. We consider here two situations: either A_i is a finite set for all i or A_i is a continuous set for all i. The utility function of each agent i is some function U_i defined on the set of joint actions $\boldsymbol{A} = A_1 \times \cdots \times A_N$, i.e., $U_i : \boldsymbol{A} \to \mathbb{R}$.

Definition 2.1.3 A game $\Gamma = (N, \{A_i\}, \{U_i\})$ is called to be *potential game* (is also denoted by the tuple $(N, \{A_i\}, \{U_i\}, \phi)$), with a potential function $\phi : \boldsymbol{A} \to \mathbb{R}$, if for any $i \in [N]$, $a^i, a'^i \in A_i$, $a^{-i} \in \times_{j \neq i} A_j$ the following holds:

$$U_i(a^i, a^{-i}) - U_i(a'^i, a^{-i}) = \phi(a^i, a^{-i}) - \phi(a'^i, a^{-i}). \tag{2.2}$$

Note that in the case of smooth utility functions U_i, $i \in [N]$, the condition (2.2) is equivalent to the following one:

$$\frac{\partial \phi(\boldsymbol{a})}{\partial a^i} = \frac{\partial U_i(\boldsymbol{a})}{\partial a^i}, \tag{2.3}$$

for any $i \in [N]$ and $\boldsymbol{a} \in \boldsymbol{A}$.

Thus, a game is potential, if there is a function that quantifies the difference in each utility function due to unilaterally deviating of the corresponding player.

In [MS96b] it was proven that if some pure strategy discrete action game is potential, then so is its extension to mixed strategies.

Proposition 2.1.1 *If some game* $\Gamma = (N, \{A_i\}, \{U_i\}, \phi)$ *with discrete action sets* $\{A_i\}$ *is potential with a potential function* ϕ, *then its extension to mixed strategies, the game* $\tilde{\Gamma} = (N, \{\Sigma_i(A_i)\}, \{\tilde{U}_i\}, \tilde{\phi})$, *is also potential with the potential function* $\tilde{\phi}$: $\tilde{\phi}(\sigma) = \sum_{\boldsymbol{a}} \phi(\boldsymbol{a})\sigma(\boldsymbol{a})$, *where* $\sigma(\boldsymbol{a}) = \sigma^1(a^1) \cdot \ldots \cdot \sigma^N(a^N)$.

In the case of continuous action potential games the result analogous to one above can be obtained.

Proposition 2.1.2 *If some game* $\Gamma = (N, \{A_i\}, \{U_i\}, \phi)$ *with continuous action sets* $\{A_i\}$, $i \in [N]$, *and utility functions* $U_i : \boldsymbol{A} \to \mathbb{R}$ *is potential with a potential function* ϕ, *then its extension to mixed strategies* $\tilde{\Gamma} = (N, \{\Sigma_i(A_i)\}, \{\tilde{U}_i\}, \tilde{\phi})$ *is also potential with the potential function* $\tilde{\phi}(\sigma) = \int_{\boldsymbol{A}} \phi(\mathbf{x}) d\sigma(\mathbf{x})$, *given the existence of the corresponding Lebesgue integrals.*

Proof Since the functions $\{U_i\}$ take values in $\mathbb{R}$, one can use the definition of the Lebesgue integrals as the limit of the Darboux sums:

$$\tilde{U}_i(\sigma) = \int_{\boldsymbol{A}} U_i(\boldsymbol{a}) d\sigma(\boldsymbol{a}) = \lim_{n\to\infty} \sum_{k=1}^{n} U_i(\boldsymbol{a}(k))\sigma(\boldsymbol{A}(k)),$$

where $\boldsymbol{A} = \bigsqcup_{k=1}^{n} \boldsymbol{A}(k)$ is a disjunctive partition of $\boldsymbol{A}$ and, thus, $\sum_{k=1}^{n} \sigma(\boldsymbol{A}(k)) = 1$, $\boldsymbol{a}(k)$ is a tagged point in $\boldsymbol{A}(k)$. Let us consider another mixed strategy

$$\sigma' \in \Sigma_1(A_1) \times \cdots \times \Sigma_N(A_N) : \sigma'^j = \sigma^j \text{ for all } j \neq i.$$

Hence, using the same partition $\{\boldsymbol{A}(k)\}$, we obtain

$$\tilde{U}_i(\sigma') = \int_{\boldsymbol{A}} U_i(\boldsymbol{a}) d\sigma'(\boldsymbol{a}) = \lim_{n\to\infty} \sum_{k=1}^{n} U_i(\boldsymbol{a}(k))\sigma'(\boldsymbol{A}(k)).$$

According to Proposition 2.1.1,

$$\sum_{k=1}^{n} U_i(\boldsymbol{a}(k))\sigma'(\boldsymbol{A}(k)) - \sum_{k=1}^{n} U_i(\boldsymbol{a}(k))\sigma(\boldsymbol{A}(k))$$
$$= \sum_{k=1}^{n} \phi(\boldsymbol{a}(k))\sigma'(\boldsymbol{A}(k)) - \sum_{k=1}^{n} \phi(\boldsymbol{a}(k))\sigma(\boldsymbol{A}(k)).$$

By taking limit as $n \to \infty$ of the both sides in the equality above we conclude the proof. □

The crucial properties of any potential game are formulated by the following theorem [SBP06].

Theorem 2.1.1 ([SBP06]) *Let $\Gamma = \{N, \{A_i\}, \{U_i\}, \phi\}$ be a potential game with potential function ϕ. Let $\boldsymbol{A}^* \subset \boldsymbol{A} = A_1 \times \cdots \times A_N$ denote the set of maxima of ϕ on $\boldsymbol{A}$ (assumed non-empty). The following statements hold true: (1) If $\boldsymbol{a} \in \boldsymbol{A}^*$, then $\boldsymbol{a}$ is a Nash equilibrium in Γ. The converse, in general, is not true. But, if, in addition, $\boldsymbol{A}$ is a convex set and ϕ is a continuously differentiable function on the interior of $\boldsymbol{A}$, then: (2) If $\boldsymbol{a}$ is a Nash equilibrium in Γ, then $\boldsymbol{a}$ is a critical point of ϕ; (3) Assume that ϕ is concave on $\boldsymbol{A}$. If $\boldsymbol{a}$ is a Nash equilibrium in Γ, then $\boldsymbol{a} \in \boldsymbol{A}^*$. If ϕ is strictly concave, such an equilibrium must be unique.*

Thus, according to the theorem above, the existence of a maximum for the function ϕ on $\boldsymbol{A}$ implies the existence of a Nash equilibrium in the game Γ. To find an equilibrium it suffices to find a maximum of the function ϕ on $\boldsymbol{A}$. Moreover, if the function ϕ is a well-behaved concave one, then the Nash equilibrium is unique in Γ. Theorem 2.1.1 implies that the set of Nash equilibria in any potential game is a subset of critical points of the potential function ϕ, if $\boldsymbol{A}$ is a convex set and ϕ is a continuously differentiable function. However, it can be noticed that the function ϕ itself is not necessarily even locally maximized at a Nash equilibrium, and, conversely, a local maximum of ϕ does not necessarily correspond to a Nash equilibrium. The relation between Nash equilibria in a potential game and stationary points of the potential function in the general case of a convex set $\boldsymbol{A}$ and a differentiable function ϕ is presented in Fig. 2.2.

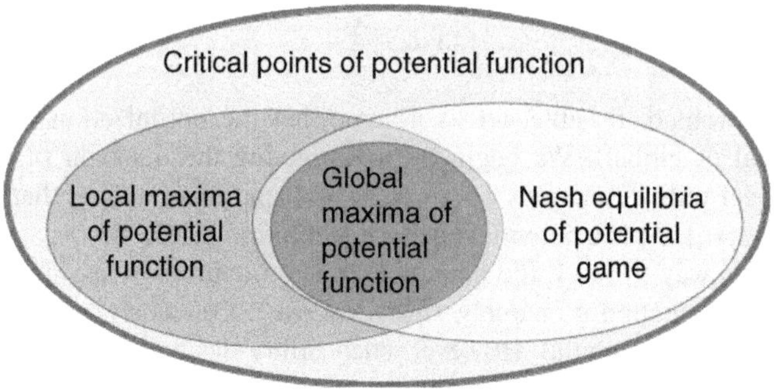

Fig. 2.2 Nash equilibria and stationary points of potential functions in potential games

2.2 Potential Game Design in Multi-Agent Optimization

The game-theoretic approach to optimization in multi-agent systems refers to the following paradigm. Firstly, the individual utility functions for agents should be modeled in such a way that certain solution concepts in the resulting game meet a system's objective. Secondly, some efficient learning rules need to be designed to learn such outcomes [MS13]. This section presents an overview of the existing methods for this approach, which have been presented in the literature so far. We emphasize here once more that this book deals with the second part of the described paradigm and its main goal is to develop some new memoryless learning procedures leading a system, that is already appropriately designed by means of potential games, to some optimal state.

2.2.1 *Multi-Agent Systems Modeled by Means of Potential Games*

Let us consider a multi-agent system consisting of N agents. Each agent $i \in [N]$ has an action set A_i, from which she can choose her action. Now let us suppose that the objective of this system is captured by some function $\phi : \boldsymbol{A} \to \mathbb{R}$, where $\boldsymbol{A} = A_1 \times \cdots \times A_N$, such that the maxima of ϕ are the system's optimal states. Thus, the common goal of the agents is to find a solution of the following optimization problem:

$$\max \phi(\boldsymbol{a}),$$

$$\text{s.t. } \boldsymbol{a} \in \boldsymbol{A}. \tag{2.4}$$

To make the problem feasible, let us assume that the optimization above has a solution (local or global). We begin by investigating the question of a possible model for local utility functions $\{U_i : \boldsymbol{A} \to \mathbb{R}\}_{i \in [N]}$ guaranteeing that the game $\Gamma = \{N, \{A_i\}, \{U_i\}, \phi\}$ is a potential one (see Definition 2.1.3).

One obvious way to reach the potential game structure is to assign each agent the same utility function ϕ, namely $U_i(\boldsymbol{a}) = \phi(\boldsymbol{a})$. Obviously, such game with *identical interests* is potential. However, such utility design is applicable only to systems with small dimensions, since in this case for calculating her utility U_i the agent i needs to get the whole information about the actions of other agents in the system. That is why in many applications some more efficient ways to model potential games were proposed. The *range restricted utilities* were demonstrated to be useful in such problems as sensor coverage [ZM13b] (will be discussed in Sect. 3.5.3) and target assignment [AMS07]. Another approach, which was initially proposed in traditional economic cost sharing literature [WT99], is called *wonderful life utilities* or *marginal contribution protocol*. This technique for utility functions' design is successfully implemented in consensus search problems [MAS09a], resource allocation problems [WT99, MS13], routing problems [Rou05], as well as in wireless network applications [Gil10].

Example 2.2.1 Marginal contributions in resource allocation.

A resource allocation problem consists in an optimal sharing of the resources from the finite set of resources $\mathcal{R}$ among N agents. Each agent $i \in [N]$ has an action set $A_i \subseteq 2^{\mathcal{R}}$ (a definite set of subsets of $\mathcal{R}$). The action set represents the set of allowable resource utilization profiles. For example, if $\mathcal{R} = \{R_1, R_2, R_3\}$, then the action set

$$A_i = \{\{R_1, R_2\}, \{R_1, R_3\}, \{R_3, R_2\}, \{R_1, R_2, R_3\}\}$$

means that the agent i can either utilize all three resources simultaneously or use only two out of three resources. Some payoff $\phi_r(\{a\}_r)$ is assigned to each resource $r \in \mathcal{R}$ in the system. This payoff depends on the set of agents who use this resource, i.e., $\{a\}_r = \{i \in [N] : r \in a^i\}$, where a^i is the chosen profile of resources from A_i. The system's global objective function in this problem is expressed by

$$\phi(a^1, \ldots, a^N) = \sum_{r \in \mathcal{R}} \phi_r(\{a\}_r) \to \max .$$

According to the marginal contribution protocol, the local utility functions of agents should be of the following type:

$$U_i(a^i, a^{-i}) = \sum_{r \in a^i} f_r^{\mathrm{MC}}(i, \{a\}_r), \text{ where } f_r^{\mathrm{MC}}(i, \{a\}_r) = \phi_r(\{a\}_r) - \phi_r(\{a\}_r \setminus \{i\}).$$

In the work [WT99] it is proven that the resulting game $\Gamma = (N, \{A_i\}, \{U_i\}, \phi)$ is potential with the potential function ϕ coinciding with the system's global objective one.

The example above deals with a multi-agent system with discrete finite states. Indeed, the action sets $\{A_i\}_{i \in [N]}$ in the example above consisted of maximum $2^{|\mathcal{R}|}$ elements (all possible resource profiles). The potential game design can, however, take place in systems with continuous actions as well.

Example 2.2.2 Potential game structure in code division multiple access (CDMA) systems.

We consider a set of mobiles $[N] = \{1, \ldots, N\}$ that share the same wireless spectrum. Each mobile $i \in [N]$ needs to transmit to a base station (multiple mobiles are allowed to transmit to the same base station). For this purpose each mobile i chooses the power allocation denoted by $p^i \geq 0$ and, thus, $\boldsymbol{p} = (p^1, \ldots, p^N)$ is the operating point of the network. For choosing some transmission power level p^i the mobile i has to pay the price $c_i(p^i)$. Let us assume that the global objective of the system is described by the following function:

$$\phi(\boldsymbol{p}) = \log\left(1 + \sum_{i \in [N]} h_i^2 p^i\right) - \sum_{i \in [N]} c_i(p^i) \to \max,$$

where h_i is the gain between the mobile i and the base station. It is proven (see, for example, [ABSA01, SBP06]) that the utility functions $\{U_i\}$ based on the users' signal to interference plus noise ratio, namely

$$U_i(\boldsymbol{p}) = \log(1 + \mathrm{SINR}_i(\boldsymbol{p})) - c_i(p^i),$$

where $\mathrm{SINR}_i(\boldsymbol{p}) = \frac{W}{B}\frac{h_i p^i}{1+\sum_{j\neq i} h_j p^j}$ and W, B are such characteristics of the CDMA system as total bandwidth and unspread bandwidth correspondingly, result in the potential game $\Gamma = (N, \{A_i\}, \{U_i\}, \phi)$, $A_i = \mathbb{R}^+, i \in [N]$. The potential function ϕ in this model is equal to the system's global objective one.

Example 2.2.3 Potential game design by means of Wonderful Life Utilities in congestion control [AB02].

Another example of potential game design in multi-agent optimization can be found in congestion control. We consider a network of $[M] = \{1, \ldots, M\}$ nodes and $[L] = \{1, \ldots, L\}$ links connecting the nodes. We assume there are $[N] = \{1, \ldots, N\}$ users in this network system. Each user $i \in [N]$ corresponds to a unique connection between the source and destination nodes $s_i, d_i \in [M]$. The route (path) R_i the user i's connection traverses is determined by her routing choice, and corresponds to a subset of $l_i \in [L]$ connecting the two nodes s_i and d_i. The nonnegative flow, $x_i \in X_i = [0, x_{i,\max}]$, sent by the user i over this path R_i. The profit the agent i gets for sending the flow x_i is defined her local payout function $u_i(x_i)$. The congestion price on the link $l \in [L]$ is defined by the function $P_l : \mathbb{R} \to \mathbb{R}$. Thus, the global system problem in this setting is defined by

$$\max_{\boldsymbol{x}=(x_1,\ldots,x_N)\in \boldsymbol{X}} \phi(\boldsymbol{x}),$$

where $\boldsymbol{X} = X_1 \times \cdots \times X_N$ and

$$\phi(\boldsymbol{x}) = \sum_{i=1}^{N} u_i(x_i) - \sum_{l\in[L]} P_l\left(\sum_{j:l\in R_j} x_j\right).$$

In the work [AB02] the technique of Wonderful Life Utilities [WT99] was used to obtain the following utility functions:

$$U_i(\boldsymbol{x}) = u_i(x_i) - \sum_{l\in R_i} P_l\left(\sum_{j:l\in R_j} x_j\right).$$

Such design provably results in the potential game $\Gamma(N, \boldsymbol{X}, \{U_i\}, \phi)$ [WT99]. A similar potential game design based on Wonderful Life Utilities can be applied, for example, to consensus problems [MS13] and broadcast tree formation problems [CK13].

2.2.2 Learning Optimal States in Potential Games

The previous subsection describes some possible methods to design local utilities of agents in a multi-agent system, in which an optimization problem (2.4) has to be solved. In many applications this design guarantees such a potential structure of the resulting game that the potential function of the game refers to the system's global objective function to be maximized. Thus, in the potential game agents (players) need to find a potential function maximizer. Now the question is, what local decision rules should be prescribed to players to allow them approaching such state.

Following standard definitions in [MS13], let us consider any learning algorithm a repeated play over discrete time $t \in \{0, 1, 2, \dots\}$.[1] At each step t, an agent $i \in [N]$ chooses her next action $a^i(t+1)$ according to the rule prescribed by the learning algorithm, namely

$$a^i(t+1) = F_i(I_i(t)) \in A_i,$$

where $I_i(t)$ is the information currently (at the step t) available to the agent i and F_i is the rule, according to which the action is updated. Note that F_i can also be a random value over A_i. Thus, to design an appropriate learning algorithm one needs firstly to define what information agents have access during the game to. In this work the following information structures in systems are discussed:

- The ideal assumption on the available information, which, unfortunately and naturally, is almost never satisfied in practical applications is *full information.* In this case each agent knows the structural form of her own utility function (it means she can estimate the function and its derivatives, if such exist, for any given argument) and is capable of observing the actions of all other agents at every step, however, not other agents' utility functions. Learning algorithms, where players do not know the utility functions of others are referred to as *uncoupled* [HMC02, HMC06]. This work is focused only on uncoupled learning dynamics. Full information learning algorithms, thus, can be generally expressed as follows:

$$a^i(t+1) = F_i(\boldsymbol{a}(0), \dots, \boldsymbol{a}(t); \{U_i^{(k)}(\cdot)\}_k),$$

  where $\boldsymbol{a}(s) = (a^1(s), \dots, a^N(s))$, as before, is the joint action vector at the moment of time s. Note that the full information assumes agents to have infinite memory, since at each step t they remember all joint actions back to the initial one.
- Another information type considered in this book is *oracle-based information with the memory of the length m.* In the case of such information structure each agent is capable of evaluating the values of her local utility associated with all

[1] One play corresponds to the iteration, where players make decisions by updating their actions.

her possible actions (also if these actions were not selected). More specifically, the learning rule for any player i can be written as follows:

$$a^i(t+1) = F_i(\{U_i(a^i, a^{-i}(t-m))\}_{a^i \in A_i}, \dots, \{U_i(a^i, a^{-i}(t))\}_{a^i \in A_i}),$$

where m is the length of the memory of the player i, namely the number of the latest utilities' values that agents memorize.[2] Even though this information assumption restricts the applicability of the corresponding algorithm in some cases, there are many settings where access to the oracle information is provided. For example, the players in routing problems can calculate their utilities simply by observing the congestion at the routes [Ros73].

- There are many situations in the applications of multi-agent systems where agents can only observe the obtained payoffs and be aware about their local chosen actions. In this case it is said that the information is *payoff-based with the memory of the length m* and the learning procedure in a system with such information should be of the following type:

$$a^i(t+1) = F_i(\{a^i(t-m), \hat{U}_i(t-m)\}, \dots, \{a^i(t), \hat{U}_i(t)\}).$$

The notation $\hat{U}_i(s)$ is used to emphasize that the agent i does not have access to the structure of U_i, but she can observe and memorize her values at each moment of time $s = t-m, \dots, t$. The specific economic literature refers to the payoff-based information as *completely uncoupled* [AB11, FY06].

- Recently cyber-physical networks [Raj15] have gained a lot of popularity in engineering. The main feature of such multi-agent systems is their large scales. Thus, to be able to make rational decisions and move toward a system optimum, agents need to exchange the information with their local neighbors. That is why the last information structure considered in this work is *communication-based information*. The learning algorithm in this case can be expressed by the following general formula:

$$a^i(t+1) = F_i\left(G(t), \{U_i^{(k)}(\cdot)\}_k\right),$$

where $G(t)$ is a communication graph in the system at the moment t, which defines information and the set of neighbors for each agent with whom this agent can exchange this information during the algorithm's run.

A lot of works are devoted to the development of the learning algorithms applicable to optimization in multi-agent systems with different information structures, given an appropriate design of potential games. *Fictitious Play* [MS96a] is an example of a learning algorithm based on the full information in the system. The main idea consists in tracking of the empirical frequency of the

[2] The current moment t is not included in the memory array.

joint actions by each agent. It is shown in [MS96a] that Fictitious Play converges to a mixed strategy Nash equilibrium in any discrete action potential game. Due to the enormous informational and computational demand of this learning procedure [GRS00, LIES05, MAS09b], some less restrictive versions of Fictitious Play are presented. As quite much information needs to be taken into account, one prefers to minimize at least memory resources required by the learning procedure [WB11, Spa12]. *Joint Strategy Fictitious Play with Inertia* [MAS09b], which is an example of algorithms using the oracle-based information with *no memory*, is proven to converge almost surely to a pure strategy Nash equilibrium in any discrete action potential game. However, since potential function maximizers represent just a subset of Nash equilibria (see Fig. 2.2), convergence to some Nash equilibrium in a potential game does not imply convergence to a system optimum. To rectify this issue a more sophisticated learning rule is proposed by the so-called *logit dynamics*. The classical (standard) logit dynamics, also known as *log-linear learning*, is initially introduced in [Blu93]. Analogously to the Joint Strategy Fictitious Play with Inertia, the information structure of the log-linear learning falls under the classification of *oracle-based information without memory* ($m = 0$). The log-linear learning has been recently applied to engineering problems in the works [MAS09a, MS12]. These works demonstrate the *stochastic stability* of potential function maximizers under the rule of the logit dynamics in discrete action games. Stochastic stability of potential function maximizers means that the learning algorithm, being applied to a multi-agent system modeled by a discrete action potential game, approaches a system optimum with the probability that can be increased by an adjustment of the algorithm's parameter. Nevertheless, this property does not imply convergence to an optimal state in any sense. This book, in its turn, provides a setting for the logit dynamics under which convergence in total variation takes place. Moreover, this work presents general properties that are sufficient in any oracle-based learning procedure without memory to guarantee convergence in total variation to a distribution over potential function maximizers in a modeled discrete action potential game. The convergence rate of such general procedure is analyzed in this book as well.

The work [MS12] studies not only the standard version of the logit dynamics with applications to discrete action potential games, but also some modifications of the learning procedure that may be applied to systems with information structures different from the oracle-based one. For example, a payoff-based version of the log-linear learning is proposed in [MS12] with the same guarantee of stochastic stability of system's optimal states (potential function maximizers). However, to implement this payoff-based algorithm agents need to have memory of the length $m = 1$. Another algorithm requiring only payoff-based information and the memory of the length $m = 1$ is *payoff-based inhomogeneous partially irrational play* [GHF12], [ZM13b]. It is shown that this learning procedure also guarantees stochastic stability of the system optima. Moreover, such a time-dependent parameter can be chosen for this algorithm that the learning procedure converges to a distribution over optima in total variation as time runs. However, all procedures mentioned above are applicable only to discrete (finite) action games. In the case of continuous actions the results presented in the work [PML15] can be used to obtain convergence to a mixed strategy logit equilibrium in systems with the oracle-based information. Roughly speaking, a logit equilibrium in a potential game is only an approximation of some critical point of the potential function [PML15]. To guarantee the stochastic stability of optimal states in a system with continuous states and oracle-based information, this book proposes the extension of both asynchronous and synchronous logit dynamics to the case of continuous action potential games.

Some other works on the game-theoretic learning over continuous actions are based on the communication protocols in the systems. For example, in the work [KNS12] a *gossip algorithm* is efficiently applied to aggregative games on graphs. The result is extended by the authors in [SP14] to the case of general continuous action games with *convex settings*. As for potential game design, the work [LM13] proposes a method to model a game with potential structure to handle optimization (minimization) problems in multi-agent systems with continuous actions and *convex objective function (costs)*. This book introduces both communication-based and payoff-based algorithms applicable to potential games with *non-concave utility functions (or, what is equivalent, non-convex costs)*. Moreover, the payoff-based algorithm can be adapted to the case of completely distributed memoryless learning of a Nash equilibrium in concave potential games. For example, such problem is often formulated for the stability analysis of electricity markets [Jen10].

In contrast to the previous work mentioned above, this book presents a generalized approach to the learning algorithms applicable to optimization in a *memoryless multi-agent system with oracle-based information and discrete actions*. This approach guarantees *convergence in total variation to a distribution over potential function maximizers* in the modeled potential game (Sect. 3.2). This approach is specified to the *standard logit dynamics and its synchronous version* (Sects. 3.3 and 3.4 respectively). Moreover, the necessity of the difference in the information structures of the asynchronous (standard logit dynamics) and synchronous log-linear learning (further called independent log-linear learning) is discussed (Sect. 3.4.1). The convergence rate estimation and the *analysis of the finite time behavior of the general memoryless algorithm* as well as of its special cases, namely of the log-linear and independent log-linear learning procedures, are provided (Sect. 3.5).

This work extends the analysis of the *standard discrete action logit dynamics to the case of continuous actions* and discusses the convergence properties of this continuous state dynamics (Sect. 3.6.1). This book presents the *continuous independent log-linear learning* and proposes a new approach to characterize the stochastic stability of potential function maximizers in this learning procedure (Sect. 3.6.2). Moreover, the analysis of *finite time behavior for continuous logit dynamics* is provided. This analysis allows setting up the algorithms' parameters efficiently (Sect. 3.6.3).

This work presents a *memoryless communication-based algorithm* leading any continuous action potential game to a local maximum of the potential function almost surely (*no concavity of utility or potential functions is assumed*) (Sect. 4.4).

For the payoff-based settings, this book proposes a *memoryless payoff-based algorithm* that converges to a local maximum of the potential function in probability, being applied to a continuous action potential game (*no concavity of utility or potential functions is assumed*) (Sect. 4.5). Furthermore, the memoryless payoff-based algorithm is adapted to be applicable to *completely distributed learning of Nash equilibria* in concave potential games (Sect. 4.6).

2.3 Distributed Optimization in Multi-Agent Systems

There are, however, situations, where optimization problems fall naturally under the game-theoretic framework and potential game design cannot be applied to solve these problems. An example of such situation is a system, where the agents' costs depend on the joint action (thus, there is still a game between agents), but the common goal is to minimize the overall cost in the system.

Example 2.3.1 Let us consider a multi-agent system in which a congestion routing problem is formulated. More specifically, there are n agents who need to transfer some amount of data from one point (start) to another (destination). Each user $i \in [n]$ gets some profit by transferring a fixed amount $x^i \in \mathbb{R}$ of her data. This profit is expressed by some function $u_i(x^i)$. On the other hand, some price $p_i(x)$ must be paid by the agent i for data transferring, where $x = \sum_{j=1}^{n} x^j$ is the overall congestion on the way between the start and the destination. Thus, the cost of the agent i is

$$c_i(\mathbf{x}) = c_i(x^1, \ldots, x^n) = p_i\left(\sum_{j=1}^{n} x^j\right) - u_i(x^i),$$

whereas the common goal of the agents is to transfer such amount of data, namely to choose such actions x^i, $i \in [n]$, that would minimize the global cost of transferring $C(\mathbf{x}) = \sum_{i=1}^{n} c_i(\mathbf{x})$.

Such problem can be considered a special case of distributed optimization in a multi-agent system. The general optimization problem in networked systems is formulated as follows. Consider a network of n agents. At each time t, the node i can only communicate to its neighbors over some communication graph. The goal of the agents is to solve the following minimization problem[3]:

$$\min_{\mathbf{z} \in Z \subseteq \mathbb{R}^d} F(\mathbf{z}) = \sum_{i=1}^{n} F_i(\mathbf{z}), \tag{2.5}$$

[3]To distinguish the framework of distributed optimization and one of the potential game designs, minimization problems are formulated in the first case in contrast to maximization problems in the second case. Moreover, note that the number of agents in distributed optimization analysis is denoted by n, whereas this number is N in potential game design.

where $F_i : \mathbb{R}^d \to \mathbb{R}$. Note that generally $d \neq n$. Problems analogous to one above often arise in the applications where large clusters of nodes need to optimize distributively some global objective (for example, the overall loss function of statistical models in big data analysis). The essential point in many applications is that the function F_i is known only to the agent i. Thus, agents need to cooperate and exchange the information about their local estimations of an optimal joint action. On the other hand, this estimation has to be consistent among agents. That is why this book is focused on an approach to distributed optimization, which is based on an *average consensus technique*. Consensus-based distributed procedures with application to the optimization problem (2.5) are originally introduced in [TBA$^+$86] and have recently gained a lot of popularity [CS12, MB13, BJ13, NO15, NO14, NRA09, RN04, RVN09, ZM13a]. The main advantage of these algorithms is that they make it possible to manage optimization in a distributed manner without a central coordinator who would need the complete information about the network's properties. Under some structural assumptions, they benefit also from simplicity of implementation and resilience to communication delays, node failures, and adversary [TLR12c, TR12, ZM14]. Moreover, the consensus-based distributed optimization techniques have been demonstrated to be useful in the stability analysis of large-scale systems [DMUH15]. The idea behind a consensus-based algorithm is the combination of a *consensus dynamics* and a *gradient descent* along the agent's local function F_i.

A lot of works are devoted to the development and analysis of distributed optimization over networks with time-invariant topology or undirected graph [GC12, DAW12, TLR12a, TLR12b, TR12]. For example, the authors in [DAW12] study a *dual averaging subgradient method* over a fixed network. In [DAW12] the connection between the spectral properties of the network and convergence rate is established. Time-varying topology is considered in the work [RNV12] with some classical requirements on a "balanced" communication setting, such as a sequence of doubly stochastic matrices corresponding to the underlying communication graph. The issue of the requirements on such "balanced" communication can be rectified by the so-called *push-sum algorithm*. The push-sum algorithm is initially introduced in [KDG03] and used in [TLR12a] for distributed optimization. The work [NO15] studies this algorithm over time-varying communication in the case of convex functions $\{F_i\}_{i=1}^n$. Moreover, the push-sum algorithm applied to convex optimization is shown to be robust to noisy measurements of gradients [NO14]. The papers [NO15, NO14] demonstrate not only convergence of this algorithm to an optimum of the convex function F, but also study its convergence rate and provide some important insights into the algorithm's performance.

Most of the theoretical work on the topic is devoted to optimization of a sum of well-behaved *convex functions*, where the assumption on subgradient existence is essentially used. However, in many applications it is crucial to have an efficient solution for *non-convex optimization* problems [MB13, SFLS14, TGL13]. In particular, resource allocation problems with non-elastic traffic are studied in [TGL13]. Such applications cannot be modeled by means of concave utility functions that render a problem a non-convex optimization (minimization), for which more sophisticated solution techniques are needed. In [MB13], the authors propose a distributed algorithm for non-convex constrained optimization based on a first order numerical method. However, convergence to local minima is guaranteed only under the assumption that the communication topology is time-invariant, the initial values of the agents are close enough to a local minimum, and a sufficiently small step-size is used. An approximate dual subgradient algorithm over time-varying network topologies is proposed in [ZM13a]. This algorithm converges to a pair of primal-dual solutions of the approximate problem given the Slater's condition for a constrained optimization problem, and under the assumption that the optimal solution set of the dual limit is singleton. Another work [BJ13] deals with the distributed projected stochastic gradient algorithm applicable to non-convex optimization. The authors demonstrate convergence of the proposed procedure to the set of Karush–Kuhn–Tucker points that represent only a necessary condition for a point to be a local minimum. Moreover, the algorithm in [BJ13] requires a double stochastic communication matrix on average, which may also restrict the range of potential applications.

Section 4.3.2 of this book demonstrates that *the push-sum distributed optimization algorithm converges almost surely to a critical point of a sum of non-convex smooth functions* under some general assumptions on the underlying functions and general connectivity assumptions on the underlying graph. A perturbation is added to the algorithm to allow the algorithm to *converge almost surely to a local optimum of this sum* (Sect. 4.3.3). *The convergence rate* of this stochastic procedure is estimated in Sect. 4.3.4.

References

[AB02] T. Alpcan, T. Basar, A game-theoretic framework for congestion control in general topology networks, in *Proceedings of the 41st IEEE Conference on Decision and Control, 2002*, vol. 2, Dec 2002, pp. 1218–1224

[AB11] I. Arieli, Y. Babichenko, Average testing and the efficient boundary. Discussion paper series. The Federmann Center for the Study of Rationality, the Hebrew University, Jerusalem (2011)

[ABSA01] T. Alpcan, T. Basar, R. Srikant, E. Altman, CDMA uplink power control as a noncooperative game, in *Proceedings of the 40th IEEE Conference on Decision and Control, 2001*, vol. 1 (2001), pp. 197–202

[AMS07] G. Arslan, J.R. Marden, J.S. Shamma, Autonomous vehicle-target assignment: a game theoretical formulation. ASME J. Dyn. Syst. Meas. Control **129**, 584–596 (2007)

[BJ13] P. Bianchi, J. Jakubowicz, Convergence of a multi-agent projected stochastic gradient algorithm for non-convex optimization. IEEE Trans. Autom. Control **58**(2), 391–405 (2013)

[Blu93] L.E. Blume, The statistical mechanics of strategic interaction. Games Econ. Behav. **5**(3), 387–424 (1993)

[CK13] F.W. Chen, J.C. Kao, Game-based broadcast over reliable and unreliable wireless links in wireless multihop networks. IEEE Trans. Mob. Comput. **12**(8), 1613–1624 (2013)

[CNGL08] E. Campos-Naòez, A. Garcia, C. Li, A game-theoretic approach to efficient power management in sensor networks. Oper. Res. **56**(3), 552–561 (2008)

[CS12] J. Chen, A.H. Sayed, Diffusion adaptation strategies for distributed optimization and learning over networks. IEEE Trans. Signal Process. **60**(8), 4289–4305 (2012)

[DAW12] J.C. Duchi, A. Agarwal, M.J. Wainwright, Dual averaging for distributed optimization: convergence analysis and network scaling. IEEE Trans. Autom. Control **57**(3), 592–606 (2012)

[DMUH15] F. Deroo, M. Meinel, M. Ulbrich, S. Hirche, Distributed stability tests for large-scale systems with limited model information. IEEE Trans. Control Netw. Syst. **3**(3), 298–309 (2015)

[FY06] D.P. Foster, H.P. Young, Regret testing: learning to play Nash equilibrium without knowing you have an opponent. Theor. Econ. **1**(3), 341–367 (2006)

[GC12] B. Gharesifard, J. Cortes, Continuous-time distributed convex optimization on weight-balanced digraphs, in *Proceedings of the 51st IEEE Conference on Decision and Control* (IEEE, New York, 2012), pp. 7451–7456

[GHF12] T. Goto, T. Hatanaka, M. Fujita, Payoff-based inhomogeneous partially irrational play for potential game theoretic cooperative control: convergence analysis, in *American Control Conference (ACC), 2012*, June 2012, pp. 2380–2387

[Gil10] R.P. Gilles, *The Cooperative Game Theory of Networks and Hierarchies*. Theory and Decision Library C (Springer, Berlin, Heidelberg, 2010)

[GRS00] A. Garcia, D. Reaume, R.L. Smith, Fictitious play for finding system optimal routings in dynamic traffic networks. Transp. Res. B Methodol. **34**(2), 147–156 (2000)

[HMC02] S. Hart, A. Mas-Colell, Uncoupled dynamics cannot lead to Nash equilibrium. Discussion paper series, The Federmann Center for the Study of Rationality, the Hebrew University, Jerusalem (2002)

[HMC06] S. Hart, A. Mas-Colell, Stochastic uncoupled dynamics and Nash equilibrium. Games Econ. Behav. **41**(1), 286–303 (2006)

[Int80] M.D. Intriligator, The applications of control theory to economics, in *Analysis and Optimization of Systems*, ed. by A. Bensoussan, J.L. Lions. Lecture Notes in Control and Information Sciences, vol. 28 (Springer, Berlin, Heidelberg, 1980), pp. 605–626

[Jen10] M.K. Jensen, Aggregative games and best-reply potentials. Econ. Theory **43**(1), 45–66 (2010)

[KDG03] D. Kempe, A. Dobra, J. Gehrke, Gossip-based computation of aggregate information, in *Proceedings. 44th Annual IEEE Symposium on Foundations of Computer Science, 2003*, Oct 2003, pp. 482–491

[KNS12] J. Koshal, A. Nedić, U.V. Shanbhag, A gossip algorithm for aggregative games on graphs, in *2012 51st IEEE Conference on Decision and Control (CDC)*, Dec 2012, pp. 4840–4845

[LIES05] T.J. Lambert III, M.A. Epelman, R.L. Smith, A fictitious play approach to large-scale optimization. Oper. Res. **53**(3), 477–489 (2005)

[LM13] N. Li, J.R. Marden, Designing games for distributed optimization. IEEE J. Sel. Top. Sign. Process. **7**(2), 230–242 (2013). Special issue on adaptation and learning over complex networks

[MAS09a] J.R. Marden, G. Arslan, J.S. Shamma, Cooperative control and potential games. Trans. Sys. Man Cyber. B **39**(6), 1393–1407 (2009)

[MAS09b] J.R. Marden, G. Arslan, J.S. Shamma, Joint strategy fictitious play with inertia for potential games. IEEE Trans. Autom. Control **54**, 208–220 (2009)

[MB13] I. Matei, J. Baras, A non-heuristic distributed algorithm for non-convex constrained optimization. Institute for Systems Research Technical Reports (2013)

[MS12] J.R. Marden, J.S. Shamma, Revisiting log-linear learning: asynchrony, completeness and payoff-based implementation. Games Econ. Behav. **75**(2), 788–808 (2012)

[MS13] J.R. Marden, J.S. Shamma, Game theory and distributed control, in *Handbook of Game Theory Volume 4*, ed. by H.P. Young, S. Zamir (Elsevier Science, Amsterdam, 2013)

[MS96a] D. Monderer, L.S. Shapley, Fictitious play property for games with identical interests. J. Econ. Theory **68**(1), 258–265 (1996)

[MS96b] D. Monderer, L.S. Shapley, Potential games. Games Econ. Behav. **14**(1), 124–143 (1996)

[Mye91] R.B. Myerson, *Game Theory, Analysis of Conflict*, 2nd edn. (Harvard University Press, Harvard, 1991)

[Nas50] J.F. Nash, Equilibrium points in n-person games. Proc. Natl. Acad. Sci. **36**, 48–49 (1950)

[NM44] J. Von Neumann, O. Morgenstern, *Theory of Games and Economic Behavior* (Princeton University Press, Princeton, 1944)

[NO14] A. Nedić, A. Olshevsky, Stochastic gradient-push for strongly convex functions on time-varying directed graphs (2014). eprint arXiv:1406.2075

[NO15] A. Nedić, A. Olshevsky, Distributed optimization over time-varying directed graphs. IEEE Trans. Autom. Control **60**(3), 601–615 (2015)

[NRA09] G. Neglia, G. Reina, S. Alouf, Distributed gradient optimization for epidemic routing: a preliminary evaluation, in *Proceedings of 2nd IFIP Wireless Days 2009*, Dec 2009

[PML15] S. Perkins, P. Mertikopoulos, D.S. Leslie, Mixed-strategy learning with continuous action sets. IEEE Trans. Autom. Control **62**, 379–384 (2015)

[Raj15] A. Rajeev, *Principles of Cyber-Physical Systems* (MIT Press, Cambridge, 2015), pp. 431–438

[RN04] M. Rabbat, R. Nowak, Distributed optimization in sensor networks, in *Proceedings of the 3rd International Symposium on Information Processing in Sensor Networks*, IPSN '04, New York, NY (ACM, New York, 2004), pp. 20–27

[RNV12] S.S. Ram, A. Nedić, V.V. Veeravalli, A new class of distributed optimization algorithms: application to regression of distributed data. Optim. Methods Softw. **27**(1), 71–88 (2012)

[Ros65] J.B. Rosen, Existence and uniqueness of equilibrium points for concave N-person games. Econometrica **33**(3), 520–534 (1965)

[Ros73] R.W. Rosenthal, A class of games possessing pure-strategy Nash equilibria. Int. J. Game Theory **2**, 65–67 (1973)

[Rou05] T. Roughgarden, *Selfish Routing and the Price of Anarchy* (The MIT Press, Cambridge, 2005)

[RVN09] S.S. Ram, V.V. Veeravalli, A. Nedić, Distributed non-autonomous power control through distributed convex optimization, in *INFOCOM* (IEEE, New York, 2009), pp. 3001–3005

[SBP06] G. Scutari, S. Barbarossa, D.P. Palomar, Potential games: a framework for vector power control problems with coupled constraints, in *2006 IEEE International Conference on Acoustics Speech and Signal Processing Proceedings*, vol. 4, May 2006, pp. 241–244

[SFLS14] G. Scutari, F. Facchinei, L. Lampariello, P. Song, Distributed methods for constrained nonconvex multi-agent optimization-part I: theory. CoRR, abs/1410.4754 (2014)

[SFS09] M. Schmalz, M. Fujita, O. Sawodny, Directed gossip algorithms, consensus problems, and stability effects of noise trading. Eur. J. Econ. Soc. Syst. **22**, 43–61 (2009)

[Sie06] T. Siegfried, *A Beautiful Math: John Nash, Game Theory, and the Modern Quest for a Code of Nature* (Joseph Henry Press, Washington, DC, 2006)

[SP14] F. Salehisadaghiani, L. Pavel, Nash equilibrium seeking by a gossip-based algorithm, in *53rd IEEE Conference on Decision and Control*, Dec 2014, pp. 1155–1160

[Spa12] M.T.J. Spaan, *Partially Observable Markov Decision Processes* (Springer, Berlin, Heidelberg, 2012), pp. 387–414

[SZPB12] W. Saad, H. Zhu, H.V. Poor, T. Basar, Game-theoretic methods for the smart grid: an overview of microgrid systems, demand-side management, and smart grid communications. IEEE Signal Process. Mag. **29**(5), 86–105 (2012)

[TA84] J.N. Tsitsiklis, M. Athans, Convergence and asymptotic agreement in distributed decision problems. IEEE Trans. Autom. Control **29**, 4250 (1984)

[TBA+86] J.N. Tsitsiklis, D.P. Bertsekas, M. Athans et al., Distributed asynchronous deterministic and stochastic gradient optimization algorithms. IEEE Trans. Autom. Control **31**(9), 803–812 (1986)

[TGL13] G. Tychogiorgos, A. Gkelias, K.K. Leung, A non-convex distributed optimization framework and its application to wireless ad-hoc networks. IEEE Trans. Wirel. Commun. **12**(9), 4286–4296 (2013)

[TLR12a] K.I. Tsianos, S. Lawlor, M.G. Rabbat, Consensus-based distributed optimization: practical issues and applications in large-scale machine learning, in *2012 IEEE 50th Annual Allerton Conference on Communication, Control, and Computing (Allerton)*, Oct 2012, pp. 1543–1550

[TLR12b] K.I. Tsianos, S. Lawlor, M.G. Rabbat, Push-sum distributed dual averaging for convex optimization, in *2012 IEEE 51st IEEE Conference on Decision and Control (CDC)*, Dec 2012, pp. 5453–5458

[TLR12c] K.I. Tsianos, S.F. Lawlor, M.G. Rabbat, Communication/computation tradeoffs in consensus-based distributed optimization, in *Advances in Neural Information Processing Systems 25: 26th Annual Conference on Neural Information Processing Systems 2012*. Proceedings of a meeting held December 3–6, 2012, Lake Tahoe, Nevada, (2012), pp. 1952–1960

[TR12] K.I. Tsianos, M.G. Rabbat, Distributed dual averaging for convex optimization under communication delays, in *American Control Conference (ACC), 2012*, June 2012, pp. 1067–1072

[WB11] P. De Wilde, G. Briscoe, Stability of evolving multiagent systems. IEEE Trans. Syst. Man Cybern. B (Cybern.) **41**(4), 1149–1157 (2011)

[WT99] D.H. Wolpert, K. Tumer, An introduction to collective intelligence. Technical report, Handbook of Agent technology. AAAI (1999)

[ZM13a] M. Zhu, S. Martinez, An approximate dual subgradient algorithm for multi-agent non-convex optimization. IEEE Trans. Autom. Control **58**(6), 1534–1539 (2013)

[ZM13b] M. Zhu, S. Martínez, Distributed coverage games for energy-aware mobile sensor networks. SIAM J. Control Optim. **51**(1), 1–27 (2013)

[ZM14] M. Zhu, S. Martinez, On attack-resilient distributed formation control in operator-vehicle networks. SIAM J. Control Optim. **52**(5), 3176–3202 (2014)

Chapter 3
Logit Dynamics in Potential Games with Memoryless Players

3.1 Introduction

This chapter deals with multi-agent systems whose objective is modeled by means of potential games and which are endowed with *oracle-based information*. As it was discussed in the previous chapter, the availability of the oracle-based information in a game assumes that each player can observe her payoff associated with any action choice given the current joint action of opponents (see Sect. 2.2.2). Thus, if we are interested here in a *memoryless learning* in a game with the oracle-based information, then the general action update rule for some player i at time t can be expressed as follows:

$$a^i(t+1) = F_i(\{a^i, U_i(a^i, a^{-i}(t))\}_{a^i \in A_i}), \tag{3.1}$$

where F_i corresponds to the algorithm's dynamics and can possibly denote some probability distribution over the action set A_i. Thus, the player i needs to be aware about the actions $\{a^i\}_{a^i \in A_i}$ she can choose from her action set as well as the realization of the utility function on these actions as a response to the currently played actions $a^{-i}(t)$ of all other players. Even though this assumption restricts the applicability of the proposed algorithm in some cases, there are many settings where the access to the oracle-based information is provided. For example, the players in routing problems can calculate their utilities simply by observing the congestion at the routes [Ros73].

Regarding the information setting above, the rule of the *logit dynamics*, which is firstly introduced in the work [Blu93], is aligned with (3.1). That is why this dynamics with its application to potential games will be in focus of the current chapter.

In its standard form the logit dynamics, also known as *log-linear learning algorithm*, runs in a game as follows. At each time step one player is randomly selected to choose an action from her action set. This choice is performed with

T. Tatarenko, *Game-Theoretic Learning and Distributed Optimization in Memoryless Multi-Agent Systems*, DOI 10.1007/978-3-319-65479-9_3

respect to some parameter β (the degree of player's rationality), and the utilities in the response to the current state of the whole system defined by the states of other players, namely $\{U_i(a^i, a^{-i}(t))\}_{a^i \in A_i}$. The rule for an action choice is constructed in such a way that a sufficiently high value of β refers to the situation when a chosen player plays the best response, i.e., chooses an action maximizing her utility function, with a high probability.

The work [Blu93] highlights that the corresponding dynamics handles two key features of strategic behavior: *lock-in* and *bounded rationality*. The lock-in property means that, once a player selects an action, she is committed to this action for some while. This fact explains why a strategy selection rule does not depend on account of the previously played actions, but only ones actually played. The bounded rationality property appears in the logit dynamics in the myopic behavior and the limited information available to players. The myopic behavior establishes that the players contemplate only the present reward and not the rewards expected in the future. Moreover, in [FL95] the *universal consistency* is suggested to be crucial for any learning algorithm. The universal consistency means that a player, updating her action according to the learning algorithm, takes at least as much utility as she could have gained had she known the frequency but not the order of observations in advance. In [FL99] this property has been proven to be fulfilled for log-linear learning. Also there exist two different interpretations of the action update rules the log-linear learning algorithm is based on. The first one corresponds to the random utility model that has been broadly adopted in Economics, whereas the second one uses the information theory concepts. For more details on these two approaches we refer the reader to [BM59] and [CT06] correspondingly. Due to all these properties, the logit dynamics can serve as an appropriate method of describing the agents' behavior in real-world applications [AMS07, MAS09a, McF74].

A significant amount of research has been devoted to studying the properties of the log-linear learning algorithm in multi-agent systems with *discrete actions*. It is well known [Blu93] that in this case the logit dynamics with some fixed parameter β defines an ergodic homogeneous Markov chain over the set of joint actions. That is why the stationary distribution exists and is unique, and the chain converges to it, independently of the initial state. Moreover, for the log-linear learning and its generalizations in discrete action potential games the states maximizing the potential function and only these states have the weights in the stationary distribution vector as $\beta \to \infty$ [Blu93, MS12]. This property concerning the log-linear learning has been studied, for example, in [LS13, MS12], and is referred to as a *stochastic stability* of the learning algorithm. The notion of stochastic stability is used to characterize which joint actions of the players have high probabilities to occur in long run of the learning procedure as the degree of players' rationality β increases.

Other works on the log-linear learning, including [AFN08, AFPP12, Ell93, MS09, You02], evaluate the time that it takes the system to achieve some specific states for some values of the rationality parameter β. In [AFPP12] the concept of logit equilibrium is introduced and the system behavior related to this state is studied for each value of β.

The distinctive feature of all works mentioned above is that they deal with long run properties of the log-linear learning in terms of a priori specified parameter β. In this case a stationary distribution of the corresponding homogeneous Markov chain estimated for a discrete action potential game [Blu93] depends on the chosen parameter β. From this dependence it follows that for each β there is some positive probability for the system to be not in the optimal states as time tends to infinity, and, thus, convergence of the system behavior to the potential function maximizers cannot be proven in such settings. For a strict theoretical analysis of convergence properties it is very important to tend the rationality parameter and time to infinity simultaneously, setting this parameter as a function $\beta(t)$ such that $\beta(t) \to \infty$ as $t \to \infty$. The analysis of the time-depending setting in the log-linear learning is one of the objectives of this chapter.

However, the classical logit dynamics is asynchronous. Indeed, only one player is chosen to update her action pro iteration. Obviously, in many applications this assumption is not fulfilled or requires a kind of a central controller guaranteeing the asynchronous behavior in the system. In the work [MS12] the authors demonstrate that the asynchrony is necessary for potential function maximizers to be stochastic stable in the log-linear learning and, thus, proposed a new algorithm still based on the logit dynamics and called *independent log-linear learning*.[1] In this algorithm agents update their actions simultaneously. Moreover, the potential function maximizers remain stochastic stable in this procedure. Nevertheless, the independent log-linear learning does not meet condition (3.1) any more, since each agent needs to observe her currently played action, namely $a^i(t)$, to update it to the next one $a^i(t+1)$. In other words, the algorithm rule has the following form:

$$a^i(t+1) = F_i(a^i(t), \{a^i, U_i(a^i, a^{-i}(t))\}_{a^i \in A_i}). \tag{3.2}$$

The extended information required by the rule above we call *supplemented oracle-based information.*

The first part of this chapter is devoted to the analysis of the *log-linear learning* algorithm and *independent log-linear learning* algorithm in *discrete action* potential games, where the algorithms' parameters are chosen to be time-dependent functions. As it has been noticed in [MS12] both learning procedures, being defined in games with discrete actions, belong to a definite class of memoryless processes, so-called regular perturbed Markov chains. That is why in Sect. 3.2 we, firstly, present the general result for regular perturbed Markov chains, claiming their strong ergodicity under some definite assumptions. The strong ergodicity of the corresponding Markov chain implies convergence in total variation of the learning algorithm to a potential function maximizer, coinciding in our model with a system's optimal state (see Sect. 2.2.1). Afterward, this result is specified for the *asynchronous log-linear learning* in Sect. 3.3. Note that this specification of the general result to

[1] Originally the algorithm is called *synchronous log-linear learning with an independent revision process*. We shorten this name to *independent log-linear learning* further on.

the asynchronous logit dynamics is consistent with the work [MRSV85], which contains convergence analysis of the centralized discrete simulated annealing procedure. To generalize the result obtained in [MS12] regarding the necessity of the asynchrony in learning procedures, Sect. 3.4 demonstrates that synchronization fails to be efficient not only in the standard logit dynamics, but also in any *oracle-based learning procedure* defined by a *regular perturbed Markov chain*. This fact justifies the introduction of *supplemented oracle-based information* into the system to set up an efficient synchronous memoryless learning algorithm. As an example of such algorithm, the *independent log-linear learning* is studied in the same Sect. 3.4. Finally, Sect. 3.5 contains the estimation of the convergence rate and the analysis of finite time behavior of the general memoryless algorithm as well as of its special cases, namely of the log-linear and independent log-linear learning procedures.

The second part of the chapter, namely Sect. 3.6, deals with memoryless oracle-based learning in potential games with *continuous actions*. The interest in continuous action games is explained by the fact that the assumption of discrete action sets does not hold in many engineering systems [MGPS07, SPB08, MBML12, SZPB12]. While the most work on the logit dynamics in games has been devoted to the case of finite actions, only few papers investigated the systems with continuous actions. Among those papers the work [PML15] should be mentioned, where the authors propose an actor-critic reinforcement learning guaranteeing convergence to a mixed strategy logit equilibrium. In distinct to this approach, this chapter extends the *log-linear learning* and *independent log-linear learning* to the case of *continuous action games in pure strategies*. In Sect. 3.6.1 *the asynchronous log-linear learning in games with continuous actions* is studied. The work analyzes long run properties of the corresponding continuous state Markov chain and demonstrates the existence of parameter settings for which the chain possesses some ergodic properties and, thus, converges in a definite sense to a potential function maximizer. It is worth noting that the analysis in this subsection is consistent with the work [HS91], where the simulated annealing process in a general state space is studied. Section 3.6.2 deals with *continuous independent log-linear learning*. Unfortunately, the Markov chain of this algorithm, even with a constant parameter, cannot be analyzed by methods applicable to the asynchronous log-linear learning in continuous action games. Although some theory has been already developed to characterize the stationary distributions of continuous state Markov chains [WY15, Fei06, New15], it seems to be a hard problem to apply this theory to the Markov chain under consideration. That is why in Sect. 3.6.2 a new approach is presented, which allows us to analyze behavior of the algorithm. The initial continuous Markov chain is approximated by chains whose long run properties can be or have been already established. As the main result, the stochastic stability of the potential function maximizers in the continuous independent log-linear learning with a constant parameter is demonstrated in the following sense. It is shown that the stationary distribution of the corresponding Markov chain places arbitrarily small probability on sets of states that are not close to ones which maximize the potential function, given an appropriate choice of the algorithm parameter. Section 3.6.3 contains the analysis of finite time behavior of the algorithm with a time-dependent parameter. This analysis turns out to be useful for the practical implementation of the learning procedure (see Sect. 3.6.4).

Remark 3.1.1 Any potential game $\Gamma = (N, \{A_i\}, \{U_i\}, \phi)$ that will be considered in this chapter is a game with bounded utility functions on $\boldsymbol{A} = A_1 \times \cdots \times A_N$. It means that utility functions $U_i : \boldsymbol{A} \to \mathbb{R}$ are such that there exists $M > 0$ such that $|U_i| \leq M$, $i \in [N]$. We assume the constant M to be known to the designer of the learning procedure in the game. That is why, without loss of generality, we can transform U_i to $U_i \in [A, B]$ for any finite A and B as follows:

$$U_i \to (B - A)\frac{U_i + M}{2M} + A \in [A, B].$$

3.2 Memoryless Learning in Discrete Action Games as a Regular Perturbed Markov Chain

As it has been mentioned in the introduction of the current chapter, the focus here is on memoryless learning procedures in potential games. These algorithms will be studied by means of the theory of Markov chains. This section, as well as Sects. 3.3–3.5, deals with discrete action games. To investigate the properties of the learning procedures in a discrete action potential game we use the fundamental results on time-homogeneous and time-inhomogeneous finite state Markov chains. These results are summarized in Appendices A.2.1 and A.2.2 correspondingly. The reader familiar with the theory of finite state Markov chains can proceed to the following subsection. Otherwise, the reader can find all necessary backgrounds in Appendix A.2.

Some results of this section are published in [Tat14b, Tat14a, Tat16b].

3.2.1 Preliminaries: Regular Perturbed Markov Chains

To prove the main results of this section we will use the theory of a specific type of Markov processes called *regular perturbed Markov chains*. That is why next the important definitions and result presented in the literature [You93] regarding these objects are described.

Let us consider a time-homogeneous, non-necessarily ergodic, Markov chain P^0 defined on some finite state space S.

Definition 3.2.1 Markov chain $P(\epsilon) = \{p_{ij}(\epsilon)\}_{i,j}$ is called a *regular perturbed Markov chain* related to the process $P^0 = \{p_{ij}^0\}_{i,j}$, if

1. $P(\epsilon)$ is ergodic for any $\epsilon > 0$,
2. $\lim_{\epsilon \to 0+} p_{ij}(\epsilon) = p_{ij}^0$ for any $i, j \in S$,
3. $p_{ij}(\epsilon) > 0$ for some $\epsilon > 0$ implies $0 < \lim_{\epsilon \to 0+} \frac{p_{ij}(\epsilon)}{\epsilon^{R_{ij}}} < \infty$ for some $R_{ij} \geq 0$.

To characterize the stationary distribution of some regular perturbed Markov chain we follow the terminology and results presented in [MS12, You93]. Consider a complete directed graph with $|S|$ vertices, one for each state. The weight on the directed edge $i \to j$ is equal to R_{ij}. A tree, T_j, rooted at vertex j, or j-tree, is a set of $|S|-1$ directed edges such that, from every vertex different from j, there is a unique directed path in the tree to j. The resistance of a rooted tree, T_j, is the sum of the weights R_{ij} on the $|S|-1$ edges that compose it. The stochastic potential of the state, j, is defined to be the minimum resistance over all trees rooted at j. With these definitions in place, the following result [MS12, You93] can be formulated:

Theorem 3.2.1 *Let $P(\epsilon)$ be a regular perturbed Markov chain related to some Markov chain P^0. Let $\boldsymbol{\pi}^\epsilon$ be the unique stationary distribution of $P(\epsilon)$ for any $\epsilon > 0$. Then $\lim_{\epsilon \to 0^+} \boldsymbol{\pi}^\epsilon = \boldsymbol{\pi}^0$, where $\boldsymbol{\pi}^0$ is some stationary distribution of P^0. Moreover, the support of the vector $\boldsymbol{\pi}^0$ contains only those states whose stochastic potential is minimal.*

3.2.2 Convergence in Total Variation of General Memoryless Learning Algorithms

Let us start by considering a generalized approach to memoryless learning algorithms in potential games. This section presents some sufficient conditions, under which such learning procedures converge in total variation to a distribution with the whole support on potential function maximizers, being applied to discrete action potential games.

A potential game $\Gamma = (N, \{A_i\}, \{U_i\}, \phi)$ is considered. The action set A_i of each agent $i \in [N]$ is discrete and finite. According to the discussion in the previous chapter, it is assumed that the set of potential function maximizers

$$A^* = \{\boldsymbol{a}^* \in \arg\max_{\boldsymbol{a} \in A} \phi(\boldsymbol{a})\}$$

is the set of optimal states in the multi-agent system modeled by the game Γ.

As we are interested in *memoryless systems*, it is assumed that the players in the game Γ are not able to retrieve the information from the previous steps. Hence, a *memoryless learning algorithm* needs to be developed in the game under consideration, in which the agents use only the current information to update their actions. Motivated by the discussion in Sect. 3.1, let every player i at each step t *independently of each other* choose an action $a_2^i \in A_i$, currently playing a_1^i, with some probability $p_{a_1^i a_2^i}(t) = p_{a_1^i a_2^i}(\epsilon(t)) \in [0, 1]$, where $\epsilon(t)$ is a time-dependent parameter of the algorithm such that $\epsilon(t) \downarrow 0$ as $t \to \infty$. Obviously, such rule renders an algorithm a time-inhomogeneous Markov chain on the joint action set

$\boldsymbol{A} = A_1 \times \cdots \times A_N$ with the transition probability matrix $P(\epsilon(t))$ defined by elements $\{p_{\boldsymbol{a}_1\boldsymbol{a}_2}(t)\}_{\boldsymbol{a}_1\boldsymbol{a}_2}$, where

$$p_{\boldsymbol{a}_1\boldsymbol{a}_2}(t) = \Pr\{\boldsymbol{a}(t+1) = \boldsymbol{a}_2 | \boldsymbol{a}(t) = \boldsymbol{a}_1\} = \prod_{i=1}^{N} p_{a_1^i a_2^i}(t).$$

We would like the learning procedure to lead the game to an optimal state coinciding with a joint action $\boldsymbol{a}^*$ from the set of potential function maximizers $\boldsymbol{A}^*$. Since the algorithm is modeled as the Markov chain $P(\epsilon(t))$, this goal corresponds to convergence in total variation of the Markov chain distribution to a distribution with the support on $\boldsymbol{A}^*$. Thus, the question is what algorithm's settings guarantee that the Markov chain $P(\epsilon(t))$ is strongly ergodic with the stationary distribution whose full support is on the set of potential function maximizers $\boldsymbol{A}^*$. The following theorem formulates some sufficient conditions under which the latter holds true.

Theorem 3.2.2 *Let the Markov chain $P(\epsilon(t))$ fulfill the following conditions:*

1. *for any fixed t, $P(\epsilon(t))$ corresponds to a regular perturbed Markov chain $P(\epsilon)$, $\epsilon = \epsilon(t)$ (see Sect. 3.2.1),*
2. *$\pi(\epsilon)$ is the unique stationary distribution of $P(\epsilon)$, $\lim_{\epsilon \to 0} \pi(\epsilon) = \pi^*$, where $\mathrm{supp}\, \pi^* = \boldsymbol{A}^*$,*
3. *for any $\boldsymbol{a}_1, \boldsymbol{a}_2 \in \boldsymbol{A}$ the transition probability $p_{\boldsymbol{a}_1\boldsymbol{a}_2}(\epsilon)$ is a rational function of ϵ.*

Then there exists such setting of the time-dependent parameter $\epsilon(t)$ that $P(\epsilon(t))$ is strongly ergodic with the stationary distribution π^*.

Proof The proof will use the result of Theorem A.2.4, Appendix A.2.2.

Firstly, we demonstrate that there exists such time-dependent parameter $\epsilon(t)$ that the time-inhomogeneous Markov chain $P(\epsilon(t))$ is weakly ergodic. According to the definition of regular perturbed Markov chains (Definition 3.2.1), the pattern of the matrices $P(\epsilon)$ is the same for all ϵ from some interval $(0, \epsilon_0)$. Then, according to Lemma A.2.1 in Appendix A.2.2, the regularity of each matrix $P(\epsilon)$ with a fixed ϵ implies the existence of such integer c, called the scrambling constant of $P(\epsilon)$, that $\prod_{j=1}^{c} P(\epsilon_j)$, $\epsilon_j \in (0, \epsilon_0)$, is a scrambling matrix. Since each $p_{\boldsymbol{a}_1\boldsymbol{a}_2}(\epsilon)$ is some rational function of ϵ and according to the definition of the coefficient of ergodicity τ (see Definition A.2.6 in Appendix A.2.2), the following holds for the time-inhomogeneous chain $P(\epsilon(t))$:

$$\alpha\left(\prod_{j=t}^{t+c-1} P(\epsilon(j))\right) = 1 - \tau\left(\prod_{j=t}^{t+c-1} P(\epsilon(j))\right) = \Theta(\epsilon(t)^d) > 0$$

for some $d > 0$. Thus, if $\epsilon(t) = \frac{1}{t^{1/d}}$, then

$$\sum_{t=0}^{\infty} \alpha\left(\prod_{j=t}^{t+c-1} P(\epsilon(j))\right) = \sum_{t=0}^{\infty} \Theta\left(\frac{1}{t}\right) = \infty.$$

Hence, according to Theorem A.2.3 in Appendix A.2.2, the Markov chain $P(\epsilon(t))$ is weakly ergodic.

Next, we show that $\sum_{t=0}^{\infty} \|\boldsymbol{\pi}(\epsilon(t+1)) - \boldsymbol{\pi}(\epsilon(t))\|_{l_1} < \infty$ for the stationary distributions $\boldsymbol{\pi}(\epsilon(t))$ of the time-homogeneous chains $\{P(\epsilon(t))\}$ for any sequence $\epsilon(t)$: $\epsilon(t) \downarrow 0$ as $t \to \infty$. Indeed, using the result presented in Theorem A.2.2 (see Appendix A.2.1) and taking into account that $p_{\boldsymbol{a}_1\boldsymbol{a}_2}(\epsilon)$ is a rational function of ϵ, we conclude that each coordinate $\pi^{\boldsymbol{a}}(\epsilon)$ in $\boldsymbol{\pi}(\epsilon)$ is a rational function of ϵ. Hence, $\pi^{\boldsymbol{a}}(\epsilon(t))$ is a rational function of t, given $\epsilon(t) = \frac{1}{t^{1/d}}$. It means that there exists $T > 0$ such that for any $\boldsymbol{a} \in \boldsymbol{A}$, $\pi^{\boldsymbol{a}}(\epsilon(t))$ is either increasing or decreasing function of t for any $t > T$. Let $\boldsymbol{A}_i$ and $\boldsymbol{A}_d$ be such subsets of $\boldsymbol{A}$, $\boldsymbol{A} = \boldsymbol{A}_i \sqcup \boldsymbol{A}_d$, that $\pi^{\boldsymbol{a}}(\epsilon(t))$, $t > T$, is an increasing function, if $\boldsymbol{a} \in \boldsymbol{A}_i$, and a decreasing function, if $\boldsymbol{a} \in \boldsymbol{A}_d$. Then, according to condition (2),

$$\begin{aligned}
\sum_{t=0}^{\infty} \|\boldsymbol{\pi}(\epsilon(t+1)) - \boldsymbol{\pi}(\epsilon(t))\|_{l_1} &= \sum_{t=0}^{T} \|\boldsymbol{\pi}(\epsilon(t+1)) - \boldsymbol{\pi}(\epsilon(t))\|_{l_1} \\
&\quad + \sum_{t=T+1}^{\infty} \sum_{\boldsymbol{a} \in \boldsymbol{A}_i} [\pi^{\boldsymbol{a}}(\epsilon(t+1)) - \pi^{\boldsymbol{a}}(\epsilon(t))] \\
&\quad + \sum_{t=T+1}^{\infty} \sum_{\boldsymbol{a} \in \boldsymbol{A}_d} [\pi^{\boldsymbol{a}}(\epsilon(t)) - \pi^{\boldsymbol{a}}(\epsilon(t+1))] \\
&= \sum_{t=0}^{T} \|\boldsymbol{\pi}(\epsilon(t+1)) - \boldsymbol{\pi}(\epsilon(t))\|_{l_1} \\
&\quad -2 \sum_{\boldsymbol{a} \in \boldsymbol{A}_i} \pi^{\boldsymbol{a}}(\epsilon(T+1)) + 2 \sum_{\boldsymbol{a} \in \boldsymbol{A}_i} \pi^{*\boldsymbol{a}} < \infty.
\end{aligned}$$

Thus, all conditions (1)–(3) of Theorem A.2.4 (see Appendix A.2.2) hold. Hence, the Markov chain $P(\epsilon(t))$ with $\epsilon(t) = \frac{1}{t^{1/d}}$ is strongly ergodic with the stationary distribution $\boldsymbol{\pi}^*$. □

We emphasize one more time that the theorem above implies the following. If some memoryless algorithm modeled by the Markov chain $P(\epsilon(t))$ fulfills the conditions of Theorem 3.2.2, then there exists the setting of the parameter $\epsilon(t)$ such that this algorithm leads the system to an optimal state. More precisely, in long run of the algorithm the probability distribution over joint actions converges in total variation to a distribution with full support on the potential function maximizers.

In the following sections it will be shown that the synchronous and asynchronous versions of the logit dynamics satisfy all conditions in Theorem 3.2.2 and, thus, can be set up to possess the desired convergence property.

3.3 Asynchronous Learning

3.3.1 *Log-Linear Learning in Discrete Action Games*

This subsection presents the discrete state dynamics of the asynchronous log-linear learning algorithm firstly introduced in [Blu93].

In some game Γ the procedure is carried out as follows:

1. At every time step t, one player $i \in [N]$ is randomly selected.
2. She updates her state choosing one action from her action set A_i according to the Boltzmann distribution with some parameter $\beta \geq 0$.

That is, the action $a^i \in A_i$ is selected with probability

$$p_{a^i}(\beta; t) = \frac{\exp\{\beta U_i(a^i, a^{-i}(t))\}}{\sum_{\hat{a}^i \in A_i} \exp\{\beta U_i(\hat{a}^i, a^{-i}(t))\}}, \tag{3.3}$$

where $a^{-i}(t) = a^{-i}(t-1)$ is the current joint action of all other players except i, i.e., $a^{-i} = (a^1, \ldots, a^{i-1}, a^{i+1}, \ldots, a^N)$. The parameter β determines how likely every player i is to select an action optimal for her, and can be interpreted as the *rationality* parameter. Indeed, from (3.3) it follows that, if $\beta = 0$, then the player i selects an action uniformly at random, but if $\beta \to \infty$, then she plays the best response (if there is more than one best response, then she chooses uniformly at random one of them).

The above dynamics defines a Markov chain with the state space equal to the set of all joint actions $\boldsymbol{A} = A_1 \times \cdots \times A_N$. From the setup of the dynamics it follows that the transition probability from the state $\boldsymbol{a}_1 \in \boldsymbol{A}$ to the state $\boldsymbol{a}_2 \in \boldsymbol{A}$ is equal to 0, if the Hamming distance[2] $d(\boldsymbol{a}_1, \boldsymbol{a}_2)$ is not less than 2, i.e., $d(\boldsymbol{a}_1, \boldsymbol{a}_2) \geq 2$, and it is $\frac{1}{N} p_{a_2^i}(\beta; t)$, if $a_1^i \neq a_2^i$, $a_1^j = a_2^j$ for all $j \neq i$. More formally, the log-linear learning can be defined as follows.

Definition 3.3.1 Let $\Gamma = (N, \{A_i\}_i, \{U_i\}_i)$ be a discrete action game with N players, players' action sets $\{A_i\}_{i=1}^N$, and utility functions $\{U_i\}_{i=1}^N$. For a chosen parameter $\beta \geq 0$ the *log-linear learning* is a Markov chain defined on the joint action set $\boldsymbol{A} = \times_{i=1}^N A_i$ with the transition probability matrix $P(\beta) = \{p_{\boldsymbol{a}_1, \boldsymbol{a}_2}(\beta)\}_{\boldsymbol{a}_1, \boldsymbol{a}_2 \in \boldsymbol{A}}$ defined as

$$p_{\boldsymbol{a}_1, \boldsymbol{a}_2}(\beta) = \frac{1}{N} \begin{cases} p_{a_2^i}(\beta), & \text{if } a_1^{-i} = a_2^{-i} \text{ and } a_1^i \neq a_2^i; \\ \sum_{i=1}^N p_{a_2^i}(\beta), & \text{if } \boldsymbol{a}_1 = \boldsymbol{a}_2; \\ 0, & \text{otherwise}; \end{cases} \tag{3.4}$$

where $p_{a_2^i}(\beta)$ is defined as in (3.3) with the omitted time parameter t.

---

[2] Hamming distance $d(\boldsymbol{a}_1, \boldsymbol{a}_2)$ between two vectors $\boldsymbol{a}_1, \boldsymbol{a}_2$ is defined as $d(\boldsymbol{a}_1, \boldsymbol{a}_2) = \sum_{a_1^i \neq a_2^i} 1$.

Note that for a constant parameter β the probability $p_{a^i}(\beta; t)$ does not depend on the moment of time t, when the player i is chosen to update her state, but only on the joint action of all other players. In other words, the probability for the chain to transit from some state $\boldsymbol{a}_1$ to another one $\boldsymbol{a}_2$ does not change with time and depends only on the coordinates of $\boldsymbol{a}_1$ and $\boldsymbol{a}_2$. It means that in the case of some constant β, the Markov chain of the log-linear learning is time-homogeneous. Such time-homogeneous Markov chain is proven to be ergodic [Blu93]. For the class of potential games the unique stationary distribution is proven to be as follows.

Theorem 3.3.1 ([Blu93]) *If $\Gamma = (N, \{A_i\}_i, \{U_i\}_i, \phi)$ is a potential game with the potential function ϕ, then the Markov chain $P(\beta)$ from Definition 3.3.1 has the unique stationary distribution $\boldsymbol{\pi}(\beta)$ with the following coordinates:*

$$\pi^{\boldsymbol{a}}(\beta) = \frac{\exp\{\beta\phi(\boldsymbol{a})\}}{\sum_{\hat{\boldsymbol{a}}\in A} \exp\{\beta\phi(\hat{\boldsymbol{a}})\}} > 0 \quad \text{for all } \boldsymbol{a} \in \boldsymbol{A}. \tag{3.5}$$

As it has been mentioned in Introduction, the reader is referred to Appendix A.2.1 for more details on the ergodicity of time-homogeneous Markov chains. Nevertheless, it is worth reminding the reader that the ergodicity of $P(\beta)$ means convergence in total variation of the chain's distribution to the stationary one $\boldsymbol{\pi}(\beta)$. It means that each joint action $\boldsymbol{a} \in \boldsymbol{A}$ is chosen by the log-linear algorithm with probability $\pi^{\boldsymbol{a}}(\beta)$ as time runs.

It is clear that as $\beta \to \infty$, all the weight of the stationary distribution $\boldsymbol{\pi}(\beta)$ is on the joint actions that maximize the potential function, i.e., on the optimal Nash equilibria. However, it does not imply the convergence of the algorithm to some optimal state, if β is chosen to be fixed. Indeed, for each chosen value of the parameter β, non-optimal outcomes occur, according to the stationary distribution $\boldsymbol{\pi}(\beta)$ in (3.5), with some positive probability. To overcome this problem one needs to consider an inhomogeneous version of the log-linear learning with a time-dependent parameter $\beta(t)$ and, hence, investigate the learning procedure by means of the time-inhomogeneous Markov chain theory (see Appendix A.2.2).

As one can guess not any setting of the parameter $\beta(t)$ implies the desired convergence. To support this guess, the next subsection provides an example in which the inhomogeneous log-linear learning with a polynomial parameter does not guarantee convergence of the joint actions to an optimal system state.

3.3.1.1 An Example: Log-Linear Learning for Consensus Problem

Let consider a simple system consisting of two agents $[N] = \{1, 2\}$ who intend to reach a consensus. Each player has her own action set $A_1 = \{0.5; 2\}, A_2 = \{0.5; 3\}$. Then the set of joint actions is $\boldsymbol{A} = \{(0.5, 0.5); (2, 3); (0.5, 3); (2, 0.5)\}$. The joint state $(0.5, 0.5)$ is a desired consensus point. To design a game for this problem we follow a standard approach to design the utility functions in consensus problems [MAS09a]: $U_i(a^i, a^j) = -|a^i - a^j|$, for $i = 1, 2$ and $j = 2, 1$. For our simple case the potential function in the designed game is $\phi(\boldsymbol{a}) = -|a^1 - a^2|$. The global maximum of this potential function, namely 0, corresponds to the consensus point: $\phi(\boldsymbol{a}) = 0 \Leftrightarrow a^1 = a^2$. We use the logit dynamics introduced in the previous subsection as the rules for action choices in the repeated game. In this setup, there is a suboptimal Nash equilibrium $(2, 3)$.

The goal of this subsection is to show that there is no convergence of the log-linear learning to the consensus state, if the parameter $\beta(t)$ is a polynomial function. More specifically, it will be demonstrated that, under such setting of the algorithm parameter, there will be always a positive probability for the system, once having reached a suboptimal Nash equilibrium, to return there as time tends to infinity. We denote the suboptimal Nash equilibrium $(2, 3)$ in our example by $\boldsymbol{a}_{\text{SNE}}$. Without restricting the generality let us assume that the initial system state is $\boldsymbol{a}_{\text{SNE}}$. The probability to return to this state $\lim_{k\to\infty} \Pr\{X_k = \boldsymbol{a}_{\text{SNE}}|X_0 = \boldsymbol{a}_{\text{SNE}}\}$ during long run of the learning algorithm can be bounded from below as follows:

$$\lim_{k\to\infty} \Pr\{X_k = \boldsymbol{a}_{\text{SNE}}|X_0 = \boldsymbol{a}_{\text{SNE}}\} \geq \prod_{t=0}^{\infty} \Pr\{X_{t+1} = \boldsymbol{a}_{\text{SNE}}|X_t = \boldsymbol{a}_{\text{SNE}}\},$$

where the right side of the inequality represents the probability to stay in the suboptimal Nash equilibrium $\boldsymbol{a}_{\text{SNE}}$ all the time.

According to Definition 3.3.1,

$$\Pr\{X_{t+1} = \boldsymbol{a}_{\text{SNE}}|X_t = \boldsymbol{a}_{\text{SNE}}\} = \frac{1}{2}\frac{e^{-\beta(t)}}{e^{-\beta(t)} + e^{-2.5\beta(t)}} + \frac{1}{2}\frac{e^{-\beta(t)}}{e^{-\beta(t)} + e^{-1.5\beta(t)}},$$

where the first and the second terms of the sum represent the situations where correspondingly the first and the second player is chosen to select an action. Next, we show that

$$\sum_{t=0}^{\infty}\left|\ln\left(\frac{1}{2(1+e^{-1.5\beta(t)})} + \frac{1}{2(1+e^{-0.5\beta(t)})}\right)\right| < \infty, \tag{3.6}$$

which is equivalent to $\prod_{t=0}^{\infty}\left(\frac{1}{2(1+e^{-1.5\beta(t)})} + \frac{1}{2(1+e^{-0.5\beta(t)})}\right) > 0$, and implies that

$$\lim_{k\to\infty} \Pr\{X_k = \boldsymbol{a}_{\text{SNE}}|X_0 = \boldsymbol{a}_{\text{SNE}}\} > 0.$$

It is straightforward to prove convergence of the sum in (3.6) by means of d'Alembert's ratio test, that claims that

$$\sum_{n=1}^{\infty} b_n < \infty,\ b_n > 0, \text{if } \lim_{n\to\infty}\frac{b_{n+1}}{b_n} < 1.$$

Thus, in our example of the consensus problem, convergence to the consensus point in long run of the log-linear learning is not fulfilled, if the parameter $\beta(t)$ is chosen as a polynomial of t. Such behavior of the system is explained by the existence of a suboptimal equilibrium state, where the system has a positive probability to return, once having reached this state. The next subsection will formulate the condition on the parameter β in the inhomogeneous log-linear learning under which convergence to optimal Nash equilibria in potential games is guaranteed.

3.3.2 Convergence to Potential Function Maximizers

This subsection deals with convergence in total variation of the log-linear learning to the uniform distribution over potential function maximizers. To provide the setting under which this convergence takes place we use the result of Theorem 3.2.2.

Firstly, we notice that the formula (3.3) and Theorem 3.3.1 allow us to conclude that the Markov chain $P(\beta)$ corresponding to the log-linear learning is a regular perturbed Markov chain $P(\epsilon)$, if we substitute $e^{-\beta}$ by ϵ. Thus, condition (1) of Theorem 3.2.2 is fulfilled. Theorem 3.3.1, which characterizes the stationary distribution of this perturbed chain, implies that condition (2) of Theorem 3.2.2 also holds. Finally, we refer to the formula (3.3) again to conclude that the transition probabilities are represented by rational functions of $\epsilon = e^{-\beta}$. Thus, condition (3) of Theorem 3.2.2 holds as well. It allows us to conclude the existence of such optimal schedule $\epsilon(t) = e^{-\beta(t)}$ that the time-inhomogeneous process $P(\beta(t))$ corresponding to the learning procedure is strongly ergodic with the unique stationary distribution π^*: $\pi^* = \lim_{t\to\infty} \pi(\beta(t))$, where the coordinates $\{\pi^a(\beta(t))\}_a$ are defined in (3.5). It implies that the time-inhomogeneous log-linear learning (ILLL) guarantees convergence in total variation of the distribution of the joint action $\boldsymbol{a}(t)$ to the uniform distribution over the set of potential function maximizers as $t \to \infty$, given an appropriate schedule for the time-dependent parameter $\beta(t)$. The following theorem formulates a variant of such time-dependent setting of the parameter $\beta(t)$ and, thus, contains the main result of this subsection. Note that due to Remark 3.1.1, we can assume that the utility functions in the game under consideration take values in the interval $[0, 1]$.

Theorem 3.3.2 *Let* $\Gamma = (N, \{A_i\}_i, \{U_i\}_i, \phi)$, $U_i : \boldsymbol{A} \to [0, 1]$, $i \in [N]$, *be a* discrete action *potential game. Let* $c < \infty$ *be the scrambling constant of the Markov chain* $P(\beta)$. *Then the inhomogeneous log-linear learning algorithm with* $\beta(t) = \frac{\ln(t+1)}{c}$ *applied to* Γ *guarantees convergence in total variation of the joint actions to the uniform distribution over potential function maximizers, in particular,* $\lim_{t\to\infty} \Pr\{\boldsymbol{a}(t) = \boldsymbol{a}^* \in \boldsymbol{A}^*\} = 1$.

Proof According to the discussion above and Theorem 3.2.2, it suffices to show that the Markov chain $P(\beta(t))$ is weakly ergodic, given the time-dependent

$\beta(t) = \frac{\ln(t+1)}{c}$. Indeed, as c is the scrambling constant of $P(\beta)$, the coefficient of ergodicity $\tau(P_{t,c}(\beta))$ of the matrix $P_{t,c}(\beta) = \prod_{k=t}^{t+c-1} P(\beta(k))$, where each element $p_{a_1,a_2}(\beta(k))$ of the matrix $P(\beta(k))$ is defined as in (3.4), is strictly less than 1. Moreover, according to definition of the coefficient of ergodicity (Definition A.2.6 in Appendix A.2.2), the value of $\alpha(P_{t,c}(\beta)) = 1 - \tau(P_{t,c}(\beta))$ can be bounded below by the minimal positive element $p(\beta)^{\min}_{t,c}$ of the matrix $P_{t,c}(\beta)$. Since U_i takes values in $[0, 1]$ for all $i \in [N]$, we get from (3.4) the following lower boundary for any positive transition probability $p(\beta(k))_{a_1,a_2}$, $\boldsymbol{a}_1, \boldsymbol{a}_2 \in \boldsymbol{A}$:

$$p(\beta(k))_{a_1,a_2} \geq \frac{1}{N} \cdot \frac{1}{1 + Ae^{\beta(k)}},$$

where $A = \max_{i=1,\ldots,N} |A_i|$. Thus, we can bound $p(\beta)^{\min}_{t,c}$ as follows:

$$p(\beta)^{\min}_{t,c} = \sum_{b_1 \in A} \cdots \sum_{b_c \in A} p(\beta(t))_{a_1,b_1} \ldots p(\beta(t+c-1))_{b_c,a_2} \geq \frac{1}{N^c} \cdot \left(\frac{1}{1 + Ae^{\beta(t+c)}}\right)^c.$$

Hence,

$$\alpha(P_{t,c}) \geq p(\beta)^{\min}_{t,c} \geq \frac{1}{N^c} \cdot \left(\frac{1}{1 + Ae^{\beta(t+c)}}\right)^c, \tag{3.7}$$

If $\beta(t) = \frac{\ln(t+1)}{c}$, then

$$\alpha(P_{t,c}) \geq \frac{1}{N^c} \cdot \left(\frac{1}{1 + A(1 + t + c)^{\frac{1}{c}}}\right)^c = \Theta\left(\frac{1}{t}\right).$$

It implies

$$\sum_{j=0}^{\infty} \alpha(P(\beta)_{jc,c}) \geq \sum_{j=1} \Theta\left(\frac{1}{j}\right) = \infty.$$

Hence, according to Theorem A.2.3 in Appendix A.2.2, the process $P(\beta(t))$ is weakly ergodic, given $\beta(t) = \frac{\ln(t+1)}{c}$. □

This subsection demonstrates that the time-inhomogeneous log-linear learning (ILLL) is a special case of the memoryless algorithms discussed in Sect. 3.2. It means that there exists an appropriate time-dependent parameter $\beta(t)$ such that the corresponding Markov chain $P(\beta(t))$ is strongly ergodic, and convergence in total variation of the algorithm to the uniform distribution over the set $\boldsymbol{A}^*$ can be concluded. Moreover, Theorem 3.3.2 provided such setting of the time-dependent parameter $\beta(t)$, under which the desired convergence takes place. However, one can notice already here that the proposed function for the algorithm parameter is a logarithmic dependence on time, which implies a very slow increase of $\beta(t)$ and as a consequence a slow convergence rate. Section 3.5 estimates this rate and provides some useful insights into the algorithm's performance given another setting of the time-dependent parameter $\beta(t)$.

3.4 Synchronization in Memoryless Learning

3.4.1 Additional Information is Needed

According to Definition 3.3.1, in the log-linear learning algorithm each player observes the current time t and uses the information about her utility function on all responses to the current joint action of others. Section 3.1 refers to this kind of information as to *oracle-based information*. Let us assume that the oracle-based information is available for agents in the systems under consideration. Note also that the necessary information in the log-linear learning does not include the knowledge on the action that is actually played by an agent chosen to update. Thus, the dependence on the required information in the action transition of the player i at the time step t in the log-linear learning can be formally expressed as follows (see also (3.1)):

$$p_{a^i}(t) = f_i(\{a^i, U_i(a^i, a^{-i}(t))\}_{a^i \in A_i}), \tag{3.8}$$

where the function f_i is defined according to (3.3), i.e.,

$$f_i(\{a^i, U_i(a^i, a^{-i}(t))\}_{a^i \in A_i}) = \frac{\exp\{\beta U_i(a^i, a^{-i}(t))\}}{\sum_{\hat{a}^i \in A_i} \exp\{\beta U_i(\hat{a}^i, a^{-i}(t))\}}. \tag{3.9}$$

We emphasize here once more that players need no history of the game for decision making in the log-linear learning. Hence, this learning procedure can be applied to those multi-agent systems in which agents do not have memory.

The previous section introduces and analyzes the log-linear learning algorithm that converges in total variation to the distribution concentrated on potential function maximizers in potential games under an appropriate setting of the time-dependent parameter. For the algorithm execution players need to have an access to the oracle-based information defined by (3.8)–(3.9). Another important feature of this learning procedure is that only *one player is allowed to update her action at each time step.* To guarantee the fulfillment of this requirement, a kind of a central controller is needed to be placed into the system. In the routing, for example, agents usually update their actions simultaneously, eventually choosing a new route, which results in a better outcome. This subsection shows that in the designed potential game players cannot reach a potential function maximizer with probability tending to one over time, if they act synchronously and the only information available for them is the oracle-based one. We present a counterexample demonstrating that there is no *memoryless learning procedure* corresponding to a *regular perturbed process* that leads a potential game to potential function maximizers under the assumption that *only oracle-based information* is available for players.

Let us consider the following 2×2 symmetric, and therefore potential, coordination game described by the table below.

	B_1	B_2
A_1	$(1, 1)$	$(0, 0)$
A_2	$(0, 0)$	$(1, 1)$

We assume that in this game two players *act simultaneously* and have access only to the oracle-based information. It is shown in [AFN08] that simultaneous update of players' actions in the log-linear learning does not lead to Nash equilibria in the game under consideration. Moreover, we can use the reasoning analogous to one in [HMC06] to show that there exists no time-homogeneous algorithm leading the game to its Nash equilibria with probability tending to one, if the agents can retrieve only the oracle-based information of type (3.8). That is why we deal with time-inhomogeneous procedures further on. Since there is no memory in the system, any learning algorithm corresponds to a Markov chain. We would like this chain to be strongly ergodic with the stationary distribution whose full support is on the Nash equilibria. We focus here on the chains defined by regular perturbed processes. The probabilities $\Pr\{A_i|B_j\}$ and $\Pr\{B_i|A_j\}$ for the first and the second agent to choose an action A_i and B_i, if the opponent is currently playing B_j and A_j, $i, j = 1, 2$ respectively, are denoted as follows: $x_1 = \Pr\{A_1|B_1\}$, $x_2 = \Pr\{A_1|B_2\}$, $y_1 = \Pr\{B_1|A_1\}$, $y_2 = \Pr\{B_1|A_2\}$. The joint action set is $\boldsymbol{A} = \{\boldsymbol{a}_1 = (A_1, B_1), \boldsymbol{a}_2 = (A_2, B_1), \boldsymbol{a}_3 = (A_1, B_2), \boldsymbol{a}_4 = (A_2, B_2)\}$. Since the available information is the oracle-based one, the following matrix P describes the transition probabilities in the Markov chain defined by the game process, where $p_{ij} = \Pr\{\boldsymbol{a}_i \to \boldsymbol{a}_j\}$:

$$P = \begin{bmatrix} x_1y_1 & (1-x_1)y_1 & x_1(1-y_1) & (1-x_1)(1-y_1) \\ x_1y_2 & (1-x_1)y_2 & x_1(1-y_2) & (1-x_1)(1-y_2) \\ x_2y_1 & (1-x_2)y_1 & x_2(1-y_1) & (1-x_2)(1-y_1) \\ x_2y_2 & (1-x_2)y_2 & x_2(1-y_2) & (1-x_2)(1-y_2) \end{bmatrix}.$$

We assume existence of an optimal schedule, i.e., existence of some time-dependent functions $x_1(t)$, $x_2(t)$, $y_1(t)$, $y_2(t)$, for the learning process under which the time-inhomogeneous Markov chain P_t fulfills the conditions in Theorem 3.2.2 and the stationary distribution vector $\boldsymbol{\pi} = (\pi^{a_1}, \pi^{a_2}, \pi^{a_3}, \pi^{a_4})$ of P_t is such that $\pi^{a_3} = \pi^{a_2} = 0$. It would mean convergence of the underlying algorithm to some Nash equilibrium in total variation. Theorem 3.2.2 allows us to conclude that the stationary distribution of the time-inhomogeneous chain P_t is the limit of the stationary distribution of the regular perturbed homogeneous chain $P(t)$ ($P(t) = P_t$ for any fixed t) as time tends to infinity. To determine this limit we use

Theorem 3.2.1. According to the definition of the oracle-based information in (3.8) and the utility values in the game under consideration,

$$1 - x_1(t) = x_2(t) = \Theta(\epsilon(t)^{g_{1,t}(0,1)}) =: \Theta(\epsilon(t)^{g_1}),$$
$$1 - y_1(t) = y_2(t) = \Theta(\epsilon(t)^{g_{2,t}(0,1)}) =: \Theta(\epsilon(t)^{g_2}),$$

where $\epsilon(t)$ is a time-dependent perturbation, $g_{i,t}(u, v)$ is the function of two variables defined by the time-dependent learning dynamics of the agent i with the oracle-based information as in (3.8). Two variables of this function correspond to the values of the utility function U_i that the agent i can observe according to the definition of the game and the oracle-based information. According to our assumption, the learning procedure under consideration converges to the Nash equilibria, and, hence, $x_1(t) = \Theta(1)$, $y_1(t) = \Theta(1)$, $1 - x_2(t) = \Theta(1)$, and $1 - y_2(t) = \Theta(1)$ as $t \to \infty$. Thus, the transition probability matrix of the regular perturbed homogeneous chain $P(t)$ is of the following form:

$$P(t) = \begin{bmatrix} \Theta(1) & \Theta(\epsilon(t)^{g_1}) & \Theta(\epsilon(t)^{g_2}) & \Theta(\epsilon(t)^{g_1+g_2}) \\ \Theta(\epsilon(t)^{g_2}) & \Theta(\epsilon(t)^{g_1+g_2}) & \Theta(1) & \Theta(\epsilon(t)^{g_1}) \\ \Theta(\epsilon(t)^{g_1}) & \Theta(1) & \Theta(\epsilon(t)^{g_1+g_2}) & \Theta(\epsilon(t)^{g_2}) \\ \Theta(\epsilon(t)^{g_1+g_2}) & \Theta(\epsilon(t)^{g_2}) & \Theta(\epsilon(t)^{g_1}) & \Theta(1) \end{bmatrix}.$$

Using the definitions regarding the theory of resistance trees presented in Sect. 3.2.1, one can check that all states of the chain $P(t)$ have the same stochastic potential. Thus, according to Theorem 3.2.1, all joint actions are contained in the support of the stationary distribution π that contradicts our assumption that $\pi^{a_3} = \pi^{a_2} = 0$. Thus, the learning procedure does not converge to the set of Nash equilibria in total variation in the proposed setting.

The next subsection demonstrates that an additional assumption on the information can be introduced to guarantee the existence of a learning algorithm, which is based on a regular perturbed procedure, following which players act synchronously and their joint actions converge to an optimal Nash equilibrium in total variation as time runs.

3.4.2 *Independent Log-Linear Learning in Discrete Action Games*

Further, in addition to the oracle-based information, each agent is allowed to observe her currently played action. Thus, the learning dynamics in the system can be set up according to the following general rule:

$$P_i(t) = f_i(a^i(t), \{a^i, U_i(a^i, a^{-i}(t))\}_{a^i \in A_i}). \tag{3.10}$$

We refer this dynamics as to the dynamics with *supplemented oracle-based information*. This subsection presents efficient rules for *simultaneous* updates of agents' actions. The corresponding algorithm can be applied to systems with the supplemented oracle-based information and converges to optimal Nash equilibria in total variation in the modeled potential games under an appropriate choice of the specific algorithm parameter.

As before, a discrete action potential game $\Gamma = (N, \{A_i\}, \{U_i\}, \phi)$ is considered. The functions U_i, $i = 1, \ldots, N$, and ϕ are defined on the set of joint actions $\boldsymbol{A}$ and take values in the interval [0, 1] (see Remark 3.1.1). The maximizers of $\phi(\boldsymbol{a})$ correspond to the system optimal states. This subsection introduces the procedure, called *independent log-linear learning*, in which agents act simultaneously and use the supplemented oracle-based information (3.10). It means that the agents not only have access to the oracle-based information, but they also can observe their currently played actions. We will demonstrate convergence of this algorithm to a distribution concentrated on potential function maximizers in total variation under some appropriate setting of the algorithm parameters.

The independent log-linear learning is firstly introduced in [MS12] to be applied to discrete action potential games. More precisely, the algorithm in the discrete action game $\Gamma = (N, \{A_i\}, \{U_i\}, \phi)$ is defined by the following rule. At each moment of time t every player i, $i \in [N]$, independently of each other updates her action $a^i(t)$ according to the probability distribution below:

$$a^i(t) = \begin{cases} a^i(t-1) & \text{w.p. } 1 - \epsilon_0^m, \\ a^i \in A_i & \text{w.p. } \epsilon_0^m \frac{\exp\{\beta U_i(a^i, \boldsymbol{a}^{-i}(t-1))\}}{\sum_{\hat{a}^i \in A_i} \exp\{\beta U_i(\hat{a}^i, \boldsymbol{a}^{-i}(t-1))\}}, \end{cases} \tag{3.11}$$

where m is some constant, β and $\epsilon_0 < 1$ are the rationality and exploration parameters correspondingly. In this algorithm agents act synchronously and do not use the information about the previous time steps. However, the supplemented oracle-based information is needed to execute the algorithm, since each agent has to know her current action in case she decides not to explore, which happens with probability $1 - \epsilon_0^m$.

Let $\beta = -\ln \epsilon_0$. Obviously, the algorithm (3.11) defines a Markov chain $P(\epsilon_0)$ on the set of joint actions $\boldsymbol{A} = A_1 \times \cdots \times A_N$. For any $\boldsymbol{a}_1, \boldsymbol{a}_2 \in \boldsymbol{A}$ the transition probability $p^{\epsilon_0}_{\boldsymbol{a}_1,\boldsymbol{a}_2}$ of this chain can be expressed as follows:

$$p_{\boldsymbol{a}_1,\boldsymbol{a}_2}(\epsilon_0) = \sum_{\mathfrak{L} \subseteq [N]: \mathfrak{I} \subseteq \mathfrak{L}} \epsilon_0^{m|\mathfrak{L}|}(1-\epsilon_0^m)^{|[N] \setminus \mathfrak{L}|} \prod_{j \in \mathfrak{L}} \frac{\epsilon_0^{-U_j(a_2^j, \boldsymbol{a}_1^{-j})}}{\sum_{a^j \in A_j} \epsilon_0^{-U_j(a^j, \boldsymbol{a}_1^{-j})}}, \tag{3.12}$$

where $\mathfrak{I}$ denotes the set $\{i : a_1^i \neq a_2^i\}$. According to these transition probabilities, the Markov chain is irreducible and aperiodic given some fixed parameters ϵ_0 and $\beta = -\ln \epsilon_0$. Thus, the chain is ergodic and has a unique stationary distribution (see Appendix A.2.1). Unfortunately, in contrast to the log-linear learning (see

Theorem 3.3.1), there is no closed form expression for the stationary distribution of the independent log-linear learning. However, the authors in [MS12] analyze long run properties of this procedure with some fixed parameters ϵ_0 and $\beta = -\ln \epsilon_0$ in the case of discrete action potential games by means of resistance trees technique. The main result obtained in that work is the proven stochastic stability of the potential function maximizers. The strict formulation of this result is provided by the following theorem.

Theorem 3.4.1 ([MS12]) *Consider any potential game* $\Gamma = (N, \{A_i\}, \{U_i\}, \phi)$ *with discrete actions such that* $U_i : \mathbf{A} \to [0, 1]$, $i \in [N]$. *The Markov chain defined on the space* $\mathbf{A} = A_1 \times \cdots \times A_N$ *by the independent log-linear learning is ergodic for any fixed parameter* ϵ_0, *given* $\beta = -\ln \epsilon_0$. *Moreover, if* $\boldsymbol{\pi}(\epsilon_0)$ *is its stationary distribution vector and* $m \geq N^2$, *then* $\lim_{\epsilon_0 \to 0} \boldsymbol{\pi}(\epsilon_0) = \boldsymbol{\pi}^*$, *where the support of* $\boldsymbol{\pi}^*$ *is on the set of potential function maximizers* $\mathbf{A}^* = \{\arg\max_{\boldsymbol{a} \in \mathbf{A}} \phi(\boldsymbol{a})\}$.

The theorem above implies that the algorithm with an appropriately chosen small exploration parameter ϵ_0 and rationality parameter $\beta = -\ln \epsilon_0$ approaches an optimal Nash equilibrium. It means that the probability to choose a potential function maximizer in long run of the algorithm can be specified by the choice of the parameters ϵ_0 and $\beta = -\ln \epsilon_0$. A smaller value of ϵ_0 corresponds to a higher probability of potential function maximizers in the stationary distribution $\boldsymbol{\pi}(\epsilon_0)$. However, according to the general formula for stationary distributions provided in Theorem A.2.2 (see Appendix A.2.1), this probability cannot be equal to 1, given a fixed parameter ϵ_0. That is why the following subsection analyzes the time-inhomogeneous version of the independent log-linear learning rendered by a time-dependent choice of the exploration parameter.

3.4.3 Convergence to Potential Function Maximizers

Now we deal with the following settings in the independent log-linear learning. Let $\epsilon_0(t) > 0$ be a positive time-dependent function decreasing to 0 as $t \to \infty$, $\epsilon(t) = \epsilon_0^m(t)$, $m = N^2$, and $\beta(t) = -\ln \epsilon_0(t)$. Then the inhomogeneous independent log-linear learning (IIndLLL) runs as follows. Each agent is endowed with the exploration parameter $\epsilon(t)$: $\lim_{t\to\infty} \epsilon(t) = 0$. At time step $t \in \mathbb{Z}_+$, each agent i chooses her next action $a^i(t+1) = a^i \in A_i$ according to the following probability distribution:

$$a^i(t) = \begin{cases} a^i(t-1), & \text{w.p. } 1 - \epsilon(t); \\ a^i \in A_i, & \text{w.p. } \epsilon(t) \dfrac{\epsilon(t)^{-\frac{U_i(a^i, a^{-i}(t))}{N^2}}}{\sum_{\hat{a}^i \in A_i} \epsilon(t)^{-\frac{U_i(\hat{a}^i, a^{-i}(t))}{N^2}}}. \end{cases} \tag{3.13}$$

Let $P(\epsilon(t))$ be a *time-inhomogeneous* Markov chain corresponding to the IIndLLL in some potential game $\Gamma = (N, \{A_i\}, \{U_i\}, \phi)$ with $U_i : \mathbf{A} \to [0, 1]$.

According to Theorem 3.4.1, $P(\epsilon(t))$ with any fixed t defines an ergodic homogeneous Markov chain such that, if $\epsilon(t) \to 0$, the stationary distribution of $P(\epsilon(t))$ has the support on the set of potential function maximizers. Moreover, from (3.12) and (3.13) it follows that the Markov chain $P(\epsilon(t))$ is a regular perturbed process with the perturbation $\epsilon(t)$ and its transition probabilities are rational functions of $\epsilon(t)$. Thus, $P(\epsilon(t))$, analogously to the Markov chain of the log-linear learning $P(\beta(t))$, fulfills the conditions of Theorem 3.2.2. Hence, there exists such setting for $\epsilon(t)$ in the IIndLLL that the corresponding Markov chain is strongly ergodic with the unique stationary distribution concentrated on the set of potential function maximizers. Recall that the strong ergodicity of $P(\epsilon(t))$ with such stationary distribution is equivalent to convergence in total variation of the algorithm's distribution to a distribution with full support on potential function maximizers. The next theorem presents some sufficient condition on $\epsilon(t)$ under which this convergence takes place.

Theorem 3.4.2 *Let* $\Gamma = (N, \{A_i\}_i, \{U_i\}_i, \phi)$, $U_i : \boldsymbol{A} \to [0, 1]$, $i \in [N]$, *be a* discrete action *potential game. Then the inhomogeneous independent log-linear learning algorithm* (3.13) *with* $\epsilon(t) = (t+1)^{-\frac{N}{N^2+1}}$ *applied to* Γ *guarantees convergence in total variation of the joint actions' distribution to a distribution with full support on the set of potential function maximizers, namely* $\lim_{t\to\infty} \Pr\{\boldsymbol{a}(t) = \boldsymbol{a}^* \in \boldsymbol{A}^*\} = 1$.

Proof According to Theorem 3.2.2, it suffices to show that under the choice

$$\epsilon(t) = (t+1)^{-\frac{N}{N^2+1}}$$

the inhomogeneous Markov chain $P(\epsilon(t))$ is weakly ergodic. Let $\alpha(P(\epsilon(t))) = 1 - \tau(P(\epsilon(t)))$, where $\tau(P(\epsilon(t)))$ is the coefficient of ergodicity of the matrix $P(\epsilon(t))$. From Definition A.2.6 and (3.13) it follows that $\alpha(P(\epsilon(t)))$ can be bounded below by the minimal positive element of $P(\epsilon(t))$. Hence,

$$\alpha(P(\epsilon(t))) \geq \min_{\boldsymbol{a}_1, \boldsymbol{a}_2 \in \boldsymbol{A}} p_{\boldsymbol{a}_1,\boldsymbol{a}_2}(\epsilon_0(t)), \tag{3.14}$$

where $p_{\boldsymbol{a}_1,\boldsymbol{a}_2}(\epsilon_0(t))$ is defined in (3.12), $\epsilon_0 = \epsilon^{\frac{1}{m}}(t)$, and $m = N^2$. Taking into account (3.12), we obtain that

$$p_{\boldsymbol{a}_1,\boldsymbol{a}_2}(\epsilon_0(t)) \geq \epsilon(t)^{\frac{N^2+1}{N}} \quad \text{for any } \boldsymbol{a}_1, \boldsymbol{a}_2 \in \boldsymbol{A}.$$

Hence, if $\epsilon(t) = (t+1)^{-\frac{N}{N^2+1}}$, (3.14) implies

$$\alpha(P(\epsilon(t))) \geq \frac{1}{t+1},$$

and, therefore,

$$\sum_{t=0}^{\infty} \alpha(P(\epsilon(t))) \geq \sum_{t=0}^{\infty} \frac{1}{t+1} = \infty.$$

Thus, we can use the criterion in Theorem A.2.3 (see Appendix A.2.2) to conclude that the time-inhomogeneous Markov chain $P(\epsilon(t))$ is weakly ergodic. □

The theorem above implies that the IIndLLL can be set up in such a way that in long run of the algorithm a potential function maximizer is chosen with probability 1.

The following section provides the estimation of the convergence rate of the IIndLLL and present the analysis of finite time behavior of the algorithm with another parameter setting. The purpose of this analysis will be, firstly, to show that in games with large dimensions the convergence rate can be slow and, secondly, to demonstrate that a faster scheduling for the parameter $\epsilon(t)$ can be chosen to guarantee convergence of the algorithm in finite time to an optimal joint action with some high given probability p that is, however, less than 1.

3.5 Convergence Rate Estimation and Finite Time Behavior

The previous sections study two special cases, the asynchronous log-linear and independent log-linear learning, of the memoryless algorithms leading a potential game to an optimal Nash equilibrium, namely to a potential function maximizer. It demonstrates existence of settings for the algorithm parameters under which the probability to choose a potential function maximizer tends to one as time runs. Next, the convergence rate of the general learning algorithm with the properties formulated in Theorem 3.2.2 is estimated. As a consequence, this general estimation can be specified to a particular case of the logit dynamics.

It is worth noting that the analysis presented further is generalization of argumentation provided in [MRSV85], where finite time behavior of the simulated annealing procedure is studied. Firstly, the following useful lemma is formulated.

Lemma 3.5.1 *For any $l \geq 1$ and any $a \in (0, 1)$ the following inequalities are fulfilled: $(l+1)^a - l^a \leq al^{a-1}$, $l^a - (l-1)^a \geq al^{a-1}$.*

Proof According to Newton's generalized binomial theorem, for $l \geq 1$, $a \in (0, 1)$, $\left(1+\frac{1}{l}\right)^a \leq 1 + \frac{a}{l}$, $\left(1-\frac{1}{l}\right)^a \leq 1 - \frac{a}{l}$. These inequalities imply the inequalities in the lemma. □

Now the focus is on the time-inhomogeneous Markov chain $P(\epsilon(t))$ figuring in Theorem 3.2.2. Let $P_{t,m} = \prod_{j=t}^{t+m-1} P(\epsilon(j))$. We omit the index ϵ in $P_{t,m}$ to simplify notations. Let $\boldsymbol{\pi}_m$ denote the state of $P(\epsilon(t))$ defined by the learning procedure discussed above at the time step $m < \infty$. It means that $\boldsymbol{\pi}_m = \boldsymbol{\pi}_0 P_{0,m}$, where $\boldsymbol{\pi}_0$ is an initial distribution of the Markov chain. Now we can write

$$\boldsymbol{\pi}_m - \boldsymbol{\pi}^* = (\boldsymbol{\pi}_m - \boldsymbol{\pi}(0)P_{0,m}) + (\boldsymbol{\pi}(0)P_{0,m} - \boldsymbol{\pi}(m)) + (\boldsymbol{\pi}(m) - \boldsymbol{\pi}^*).$$

Here, similar to Theorem 3.2.2, $\pi(t) = \pi(\epsilon(t))$ is the stationary distribution of the time-homogeneous Markov chain $P(\epsilon(t))$ with any fixed t and π^* is the stationary distribution of the time-inhomogeneous chain $P(\epsilon(t))$. Thus,

$$\|\pi_m - \pi^*\|_{l_1} \le \|\pi_m - \pi(0)P_{0,m}\|_{l_1} + \|\pi(0)P_{0,m} - \pi(m)\|_{l_1} + \|\pi(m) - \pi^*\|_{l_1}. \tag{3.15}$$

We proceed to estimate the convergence rate by bounding each term on the right-hand side of the inequality in (3.15).

First Term Let c be the scrambling constant of $P(\epsilon(t))$, i.e., c is the minimal constant such that the matrix $\prod_{j=t}^{t+c-1} P(\epsilon(j))$ is scrambling for any t. Using Lemma A.2.2 in Appendix A.2.2, we obtain the following estimation of the first term in (3.15):

$$\|\pi_{mc} - \pi(0)P_{0,mc}\|_{l_1} = \|\pi_0 P_{0,mc} - \pi(0)P_{0,mc}\|_{l_1} \le \|\pi_0 - \pi(0)\|_{l_1} \tau(P_{0,mc}). \tag{3.16}$$

To complete the estimation of the first term in (3.15), we prove the following proposition allowing us to bound the coefficient of ergodicity $\tau(P_{0,mc})$.

Proposition 3.5.1 *Let $P(\epsilon(t))$ be a regular perturbed process that defines a learning procedure with the parameter $\epsilon(t)$ and satisfies the conditions in Theorem 3.2.2. Then there exist constants $l \ge 1$ and $L \in (0, 1)$ such that $\tau(P_{0,mc}) \le \left(\frac{l}{m+1}\right)^L$.*

Proof Let $\tau(nc - c, mc - 1) := \tau(P_{nc-c,mc-nc+c})$, $m \ge n \ge 1$. Using the reasoning analogues to the proof of Theorem 3.2.2, we obtain

$$\tau(kc - c, kc - 1) \le 1 - \gamma(kc) \text{ for any } k \ge 1,$$

where $\gamma(kc) = \alpha(\prod_{j=kc}^{kc+c-1} P(\epsilon(j))) = \Theta(1/kc)$, $\gamma(kc) \in (0, 1)$, given an appropriate choice of $\epsilon(t)$. The equality $\gamma(kc) = \Theta(1/kc)$ implies that there exists such l that $\frac{L'}{kc} \le \gamma(kc) < 1$ for any $k \ge l$ and some $L' \in (0, 1)$. Using Lemma A.2.3 in Appendix A.2.2, we obtain

$$\tau(lc - c, mc - 1) \le \prod_{k=l}^{m} \tau(kc - c, kc - 1) \le \prod_{k=l}^{m} (1 - \gamma(kc)).$$

Thus, taking into account that $\gamma(kc) \in (0, 1)$ and given $L = L'/c$, we get

$$\tau(0, mc - 1) \le \tau(lc - c, mc - 1) \le \prod_{k=1}^{m} \left(1 - \frac{L}{k}\right)$$
$$\le \exp\left\{\sum_{k=l}^{m} \ln\left(1 - \frac{L}{k}\right)\right\} \le \exp\left\{L \cdot \sum_{k=l}^{m} -\frac{1}{k}\right\} \le \left(\frac{l}{m+1}\right)^L, \tag{3.17}$$

where the last two inequalities are due to the inequalities $\ln(1-y) \le -y$ for $y \in [0, 1]$ and $e^{-x} \le \frac{1}{1+x}$ for $x \ge 0$. □

Second Term It is shown in [MRSV85] that, for any time-inhomogeneous Markov chain P_t whose transition probability matrix for any fixed t is ergodic with the stationary distribution $\pi(t)$ and the scrambling constant is c, the following holds:

$$\boldsymbol{\pi}(0)P_{0,mc} - \boldsymbol{\pi}(mc) = \sum_{k=1}^{m} \theta(kc)P_{kc-c,mc-kc+c},$$

$$\theta(kc) = \sum_{j=0}^{c} (\boldsymbol{\pi}(kc-j) - \boldsymbol{\pi}(kc-j+1))P_{kc-j+1,j}.$$

As we are interested in asymptotic of the distribution $\boldsymbol{\pi}_{mc}$, we can assume that there exists such $m' < m$ that for any $t > m'$ and any $\boldsymbol{a} \in \boldsymbol{A}$ the coordinate $\pi^{\boldsymbol{a}}(t)$ is either an increasing or a decreasing function of t. Recall that $\boldsymbol{A}_i$ is the subset of $\boldsymbol{A}$ such that for any $\boldsymbol{a} \in \boldsymbol{A}_i$ the coordinate $\pi^{\boldsymbol{a}}(t)$ is an increasing function of t, if $t \ge T$ (in terms of the proof of Theorem 3.2.2, $m' = T$). Then, applying Lemma A.2.3 from Appendix A.2.2 and then (3.17), we obtain

$$\begin{aligned}\|\boldsymbol{\pi}(0)P_{0,mc} - \boldsymbol{\pi}(mc)\|_{l_1} &\le \sum_{k=1}^{m} \tau(kc-c, mc-1) \sum_{j=0}^{c} \|\boldsymbol{\pi}(kc-j) - \boldsymbol{\pi}(kc-j+1)\|_{l_1} \\ &\le \frac{C}{(m+1)^L} + \frac{2}{(m+1)^L} \sum_{k=m'}^{m} (k+1)^L(\pi^*(kc+1) \\ &\qquad -\pi^*(kc-c)),\end{aligned} \tag{3.18}$$

for some $C > 0$, where $\pi^*(t) = \sum_{a \in A_i} \pi^a(t)$, and last inequality is due to the fact that $\|\boldsymbol{\pi}(t+1) - \boldsymbol{\pi}(t)\|_{l_1} = 2(\pi^*(t+1) - \pi^*(t))$ for $t \ge m'$. Let us introduce $\pi^*_-(t) = 1 - \pi^*(t)$. Then,

$$\begin{aligned}&\sum_{k=m'}^{m} (k+1)^L(\pi^*(kc+1) - \pi^*(kc-c)) \le \left(\sum_{k=m'}^{m} ((k+1)^L - k^L)\pi^*_-(kc-c)\right) \\ &\quad + 2^L\pi^*_-(m'c-c) \le \left(\sum_{k=m'}^{m} \frac{L}{k^{1-L}}\pi^*_-(kc-c)\right) + 2^L\pi^*_-(m'c-c),\end{aligned} \tag{3.19}$$

where the last inequality is due to Lemma 3.5.1.

To finish the estimation, we note that $\pi^*_-(t)$ is a rational function of t that decreases for $t \ge m'$. Therefore, there exists such constant $Q' \ge 0$ that $\pi^*_-(t) = \Theta(1/t^{Q'})$ as $t \to \infty$ and, hence, $\pi^*_-(t) \le \frac{M}{t^{Q'}}$ for any sufficiently large t. Thus, taking into account (3.18), (3.19), and given any $Q < \min(Q', L)$ and some $\hat{m}$, we conclude that

$$\|\boldsymbol{\pi}(0)P_{0,mc} - \boldsymbol{\pi}(mc)\|_{l_1} \le \frac{C'}{(m+1)^L} + \frac{2LM}{(m+1)^L}\sum_{k=\hat{m}}^{m}(k+1)^{(L-Q)-1} \le \frac{C'}{(m+1)^L}$$
$$+ \frac{2LM}{(L-Q)(m+1)^L}\sum_{k=\hat{m}}^{m}(k+1)^{L-Q} - k^{L-Q} \le \frac{C'}{(m+1)^L} + \frac{2LM}{(L-Q)(m+1)^Q}.$$

Third Term Since $m > m'$, we get

$$\|\boldsymbol{\pi}(m) - \boldsymbol{\pi}^*\|_{l_1} = 2\sum_{a\in A_i}\pi^{*a} - 2\sum_{a\in A_i}\pi^a(m) \le 2 - 2\sum_{a\in A_i}\pi^a(m) = 2\pi_-^*(m) \le \frac{2M}{m^{Q'}}.$$

All three estimations together with (3.15) can be summarized as follows. Let $P(\epsilon(t))$ be a regular perturbed Markov process for which the conditions in Theorem 3.2.2 hold and let $\epsilon(t)$ be the parameter guaranteeing strong ergodicity of the time-inhomogeneous $P(\epsilon(t))$. Moreover, let

1. c be the scrambling constant of $P(\epsilon(t))$,
2. L be such that $\lim_{t\to\infty} t\alpha(\prod_{j=ct}^{tc+c-1} P(\epsilon(j))) = L$,
3. Q' be such that $\sum_{a\in A_d}\pi^a(t) = \Theta(1/t^{Q'})$ as $t \to \infty$.

Then the theorem below can be formulated.

Theorem 3.5.1 *The convergence rate of the time-inhomogeneous Markov chain $P(\epsilon(t))$ to its unique stationary distribution $\boldsymbol{\pi}^*$ can be estimated as follows:*

$$\|\boldsymbol{\pi}_{mc} - \boldsymbol{\pi}^*\|_{l_1} = O\left(\frac{1}{m^Q}\right), \quad m \to \infty,$$

where Q is any constant less than $\min\{Q', L\}$.

The theorem above implies that the convergence rate of the learning algorithm with an optimal schedule is bounded by $\min(Q', L)$, where L characterizes the converge of the ergodicity coefficient to 0, and Q' describes the asymptotic of the coordinates in the stationary distribution $\boldsymbol{\pi}(t)$ of the time-homogeneous chain as $t \to \infty$. Both L and Q' depend on the scrambling constant, number of actions, and the range of utility functions. For large game dimensions and a weak connectivity structure of the underlying chain[3] the constant Q can occur to be very small. Unfortunately, the upper bound (3.15) does not allow us to conclude the asymptotic tightness of the convergence rate in Theorem 3.5.1 generally. However, some game example can be presented to demonstrate that the actual value of $\|\boldsymbol{\pi}_{mc} - \boldsymbol{\pi}^*\|_{l_1}$

[3]The weak connectivity structure of some ergodic Markov chain means a high power in which its transition probability matrix is scrambling.

is close to $\frac{1}{m^Q}$ as $m \to \infty$, which implies a low convergence rate of the learning algorithm (see discussion in the next subsection).

3.5.1 Convergence Rate of Time-Inhomogeneous Log-Linear Learning

To estimate the convergence rate of the time-inhomogeneous log-linear learning (ILLL) algorithm (see Sect. 3.3) with $\beta(t) = \frac{\ln(t+1)}{c}$, where c is the scrambling constant, we use the general result formulated in Theorem 3.5.1.

Recall that under this choice of $\beta(t)$ the Markov chain of the learning procedure corresponds to a regular perturbed process with the transition probabilities expressed by rational functions of t. According to (3.7),

$$\lim_{t\to\infty} \frac{\alpha(P_{t,c})}{t} \geq \frac{1}{(2AN)^c},$$

where $\alpha(P_{t,c}) = 1 - \tau(P_{t,c})$ and τ is the coefficient of ergodicity of $P_{t,c}$. Note that the argument $\beta(t)$ in $P_{t,c}$ is omitted for the notation simplicity. Thus, in the case of the log-linear learning, the constant L in Theorem 3.5.1 fulfills

$$L \geq \hat{L} = \frac{1}{c(2AN)^c}.$$

Now the constant Q' figuring in Theorem 3.5.1 will be estimated. Let, as before, $\pi(t) = \pi(\beta(t))$ be the stationary distribution of the time-homogeneous Markov chain $P(\beta(t))$ for the fixed t (see (3.5)). The formula (3.5) implies that $\pi^{a}(t)$ is an increasing function of t, i.e., $\boldsymbol{a} \in \boldsymbol{A}_i$, if and only if $\boldsymbol{a} \in \boldsymbol{A}^*$. Moreover,

$$\pi^*_-(t) = 1 - \sum_{a \in A^*} \pi^{a}(t) = \frac{\sum_{a \notin A^*} (t+1)^{\frac{\phi(a)}{c}}}{\sum_{a \in A^*} (t+1)^{\frac{\phi(a)}{c}}} = \Theta(1/t^{Q'}),$$

where $Q' = \min_{a \notin A^*} \frac{\phi^* - \phi(a)}{c}$, $\phi^* = \max_{a \in A} \phi(\boldsymbol{a})$. Thus, the following theorem is obtained.

Theorem 3.5.2 *The convergence rate of the inhomogeneous log-linear learning with* $\beta(t) = \frac{\ln(t+1)}{c}$, *where* c *is the scrambling constant of* $P(\beta(t))$ *is estimated as follows:*

$$\|\boldsymbol{\pi}_{ct} - \boldsymbol{\pi}^*\|_{l_1} = O\left(\frac{1}{t^Q}\right).$$

Here Q *is any constant less than* $\min(Q', \hat{L})$, $Q' = \min_{a \notin A^*} \frac{\phi^* - \phi(a)}{c}$, $\phi^* = \max_{a \in A} \phi(\boldsymbol{a})$, $\hat{L} = 1/c(2AN)^c$.

The theorem above describes the convergence rate of the ILLL under an appropriate choice of $\beta(t)$. We can see that convergence is polynomial in time. However, it is exponential in the parameters $\hat{L}$ and Q'. Thus, the upper bound in the convergence rate estimation increases exponentially with the decrease of the values $\hat{L} = 1/c(AN)^c$ or $Q' = \min_{a \notin A^*} \frac{\phi^* - \phi(a)}{c}$. It implies that the upper bound for the convergence rate can be very large for a game of some high dimension or in the case when the second optimal value of the potential function is close to the optimal one. As it has been mentioned above, the convergence rate estimation in Theorem 3.5.2 is expressed in terms of big O notation and, thus, is not tight. However, the example below demonstrates that this upper bound can be very close to the lower bound of the convergence rate.

Let us turn back to the example introduced in Sect. 3.3.1 of the system with two agents $[N] = \{1, 2\}$. Recall that in this game each player has her own action set $A_1 = \{0.5; 2\}$, $A_2 = \{0.5; 3\}$ and the global objective in the system is to reach a consensus. The set of joint actions is $\boldsymbol{A} = \{\boldsymbol{a}_1 = (0.5, 0.5); \boldsymbol{a}_2 = (2, 3); \boldsymbol{a}_3 = (0.5, 3); \boldsymbol{a}_4 = (2, 0.5)\}$. The joint action $\boldsymbol{a}_1 = (0.5, 0.5)$ is the unique consensus point. As before, to design a game for this problem we follow a standard approach to design the utility functions in consensus problems [MAS09a]:

$$U_i(a^i, a^j) = -|a^i - a^j|, \text{ for } i = 1, 2 \text{ and } j = 2, 1.$$

For our simple case the potential function in the designed game is $\phi(\boldsymbol{a}) = -|a^1 - a^2|$. The global maximum of this potential function, namely 0, corresponds to the consensus point: $\phi(\boldsymbol{a}) = 0 \Leftrightarrow \boldsymbol{a} = \boldsymbol{a}_1$. To get to the optimal equilibrium, the players use the log-linear learning with the optimal scheduled parameter $\beta(t) = \frac{\ln(t+1)}{2.5}$, since in this setting $c = 1$ and the factor 2.5 is used to adjust the values of utility functions to the interval [0, 1]. Let the game be initiated at $\boldsymbol{a}_2 = (2, 3)$, hence, $\boldsymbol{\pi}(0) = (0, 1, 0, 0)$. Notice that, since $\boldsymbol{\pi} = (1, 0, 0, 0)$,

$$\|\boldsymbol{\pi}(0)P_{0,t} - \boldsymbol{\pi}\|_{l_1} \geq P_{0,t}(\boldsymbol{a}_2, \boldsymbol{a}_2),$$

where $P_{0,t}(\boldsymbol{a}_2, \boldsymbol{a}_2)$ is the probability to get to the joint action $\boldsymbol{a}_2$ starting from the same $\boldsymbol{a}_2$ after t iterations. Obviously,

$$P_{0,t}(\boldsymbol{a}_2, \boldsymbol{a}_2) > \prod_{k=1}^{t} p_{\boldsymbol{a}_2, \boldsymbol{a}_2}(k) = \prod_{k=1}^{t} \frac{1}{2}\left(\frac{k^{0.2}}{k^{0.2}+1} + \frac{k^{0.6}}{k^{0.6}+1}\right) > \prod_{k=1}^{t} \frac{k^{0.2}}{k^{0.2}+1}.$$

Now, using the equality

$$\prod_{k=1}^{t} \frac{k^{0.2}}{k^{0.2}+1} = \exp\left\{\sum_{k=1}^{t} \ln\left(\frac{k^{0.2}}{k^{0.2}+1}\right)\right\}$$

and the inequality

$$\sum_{k=1}^{t} \ln\left(\frac{k^{0.2}}{k^{0.2}+1}\right) \geq -\sum_{k=1}^{t} \ln(k^{0.2}+1) \geq -\int_{0}^{t+1} \ln(x^{0.2}+1)dx \geq -\ln((t+1)^{0.2}+1),$$

we get

$$P_{0,t}(\boldsymbol{a}_2, \boldsymbol{a}_2) > \frac{1}{(t+1)^{0.2}+1}$$

and, thus,

$$\|\boldsymbol{\pi}(0)P_{0,t} - \boldsymbol{\pi}\| = \Omega\left(\frac{1}{t^{0.2}}\right).$$

On the other hand, Theorem 3.5.2 gives the estimation $\|\boldsymbol{\pi}(0)P_{0,t} - \boldsymbol{\pi}^*\| = O\left(\frac{1}{t^Q}\right)$, where Q is any constant less than 0.125.

Thus, the convergence rate of the learning algorithm based on a regular perturbed process can be very close to its upper bound established in Theorem 3.5.1. It implies a slow convergence rate of the algorithm. To accelerate this rate, in practical applications one sets up the algorithm parameter $\beta(t)$ to be a polynomial function of t. The corresponding simulations demonstrate the efficiency of this choice [AMS07, MAS09a]. However, no theoretic explanation has been given so far to support such choice in that works. Although the polynomial parameter $\beta(t)$ does not imply convergence to the set of potential function maximizers in total variation (see the example in Sect. 3.3.1), the next theorem supports the idea to set a "faster" schedule to assure the optimal action selection with a high probability after some finite number of iterations.

Theorem 3.5.3 *Let* $\Gamma = (N, \{A_i\}_i, \{U_i\}_i, \phi)$ *be a* discrete action *potential game. Let the log-linear learning with* $\beta(t) = \frac{(t+1)^n}{S}$, *where n is fixed, be applied to* Γ. *Let c be the scrambling constant of the corresponding Markov chain and* $\boldsymbol{\pi}^*$ *be the probability vector uniformly distributed over potential function maximizers. Then for any given* $\varepsilon \in (0,1)$ *there exist T and* $S = S(T)$ *such that*

$$\|\boldsymbol{\pi}_{ct} - \boldsymbol{\pi}^*\|_{l_1} \leq \varepsilon$$

for any $t \geq T$.

Proof As before, $\pi(t) = \pi(\beta(t))$ denotes the stationary distribution of the time-homogeneous Markov chain $P(\beta(t))$ for any fixed t. First of all, we notice that

$$\lim_{t\to\infty} \pi(\beta(t)) = \pi^*$$

under the setting $\beta(t) = \frac{(t+1)^n}{S}$ (see (3.5)). Hence, for any given $\varepsilon > 0$ we can find such T_1 that for any $t \geq T_1$

$$\|\pi(k+ct) - \pi^*\|_{l_1} \leq \frac{\varepsilon}{3}, \tag{3.20}$$

for any $k \geq 0$. Let us define $\epsilon(t) = e^{\beta(t)}$. Then the transition probabilities are represented by rational functions of $\epsilon(t)$ and we can use the reasoning analogous to one in Theorem 3.2.2 to demonstrate that

$$\sum_{t=1}^{\infty} \|\pi(t) - \pi(t-1)\|_{l_1} < \infty.$$

Hence, there exists T_2 such that for any given $\varepsilon > 0$ and $k > T_2$

$$\sum_{j=k}^{\infty} \|\pi(j) - \pi(j-1)\|_{l_1} \leq \frac{\varepsilon}{3}. \tag{3.21}$$

Using the triangle inequality, we can get for any $k \geq 0$:

$$\begin{aligned}\|\boldsymbol{\pi}_{ct} - \pi^*\|_{l_1} \leq \|\boldsymbol{\pi}_{ct} - \pi(k)P_{k,ct}\|_{l_1} \\ +\|\pi(k)P_{k,ct} - \pi(k+ct)\|_{l_1} + \|\pi(k+ct) - \pi^*\|_{l_1}.\end{aligned} \tag{3.22}$$

Since $\pi(k)P_{k,1} = \pi(k)P(\beta(k)) = \pi(k)$, we get

$$\begin{aligned}\pi(k)P_{k,ct} - \pi(k+ct) &= \pi(k)P_{k+1,ct-1} - \pi(k+ct) \\ &= \pi(k)P_{k+1,ct-1} - \pi(k+1)P_{k+1,ct-1} \\ &\quad +\pi(k+1)P_{k+1,ct-1} - \pi(k+ct) \\ &= \ldots = \sum_{j=k}^{k+ct-1} (\pi(j) - \pi(j+1))P_{j+1,ct-(j-k+1)}.\end{aligned} \tag{3.23}$$

Thus, according to (3.21),

$$\|\pi(k)P_{k,ct} - \pi(k+ct)\|_{l_1} \leq \sum_{j=k}^{\infty} \|(\pi(j) - \pi(j+1))P_{j+1,ct-(j-k+1)}\|_{l_1} < \frac{\varepsilon}{3} \tag{3.24}$$

for any $k > T_2$.

Since c is the scrambling constant of the system, we obtain for $\beta(t) = \frac{(t+1)^n}{S}$

$$\tau(P_{k,ct}) \leq \prod_{j=k}^{t} \left(1 - \frac{1}{(2AN)^c \exp\left(\frac{(j+c)^n}{S}\right)}\right) \leq \left(1 - \frac{1}{(2AN)^c \exp\left(\frac{(t+c)^n}{S}\right)}\right)^t.$$

As $\varepsilon \in (0,1)$ there exists T_3 such that $\left(\frac{\varepsilon}{3}\right)^{\frac{1}{t}} > 1 - \frac{1}{(2AN)^c}$ for any $t \geq T_3$. It means that $\frac{1}{(2AN)^c\left(1-(\varepsilon/3)^{\frac{1}{T}}\right)} > 1$ for $T = \max(T_1, T_3)$. It allows us to define the constant $S = S(T)$ to be the minimal constant for which $\exp\left(\frac{(T+c)^n}{S}\right) \leq \frac{1}{(2AN)^c\left(1-(\varepsilon/3)^{\frac{1}{T}}\right)}$, that is equivalent to $\left(1 - \frac{1}{(2AN)^c \exp\left(\frac{(T+c)^n}{S}\right)}\right)^T \leq \frac{\varepsilon}{3}$. Hence, $\tau(P_{k,cT}) \leq \frac{\varepsilon}{3}$. Due to Lemma A.2.3 of Appendix A.2.2, for any $t \geq T$ the previous inequality holds, namely, $\tau(P_{k,ct}) < \frac{\varepsilon}{3}$. Hence, taking into account Lemma A.2.2 of Appendix A.2.2, we obtain

$$\|\boldsymbol{\pi}_{ct} - \boldsymbol{\pi}(k)P_{k,ct}\|_{l_1} \leq \tau(P_{k,ct}) < \frac{\varepsilon}{3} \tag{3.25}$$

for any $t \geq T$ and the chosen $S = S(T)$. Thus, bringing (3.20), (3.22), (3.24), and (3.25) together, we conclude existence of T and $S(T)$ such that under the setting $\beta(t) = \frac{(t+1)^n}{S(T)}$

$$\|\boldsymbol{\pi}_{ct} - \boldsymbol{\pi}^*\|_{l_1} \leq \varepsilon$$

for any $t > T$ and given $\varepsilon > 0$. □

As we have seen in the proof above, in the case of any polynomial $\beta(t) = \frac{(t+1)^n}{S}$, n is fixed, the Markov chain $P(\beta(t))$ fulfills conditions (2) and (3) of Theorem 3.2.2. But this process is not weakly ergodic (for more details see [Tat14b]) and, thus, the system's behavior essentially depends on the initial state. Nevertheless, according to the proof of Theorem 3.5.3, for any fixed ε there exists T, for with an adjustment parameter $S(T)$ can be found, such that the distance between the Markov chain $P(\beta(t))$, $\beta(t) = \frac{(t+1)^n}{S}$, and the uniform distribution over potential function maximizers $\boldsymbol{\pi}^*$ is less than ε after T iterations.

3.5.2 *Convergence Rate of Time-Inhomogeneous Independent Log-Linear Learning*

To estimate the convergence rate of the independent log-linear learning, we refer again to the result formulated in Theorem 3.5.1. According to convergence analysis presented in Sect. 3.3, this learning algorithm has the following properties:

- $c = 1$. The definition of the independent log-linear learning algorithm implies that each transition probability matrix of the corresponding Markov chain is scrambling.
- Let the algorithm parameters be chosen as $\epsilon(t) = (t+1)^{-\frac{N}{N^2+1}}$. Then $\tau(P_{l,m}) \leq \left(\frac{l}{m+1}\right)$, for any $1 \leq l \leq m$ (see Theorems 3.4.2 and 3.5.1).

- According to (3.13) and Theorem A.2.2 (Appendix A.2.1), $\pi^*_-(t) = \sum_{a\in A_d} \pi^a(t) = \Theta(\epsilon(t)^{b'})$, as $\epsilon(t) \to 0$, where b' depends on the properties of the potential function, as well as on the number of agents, and can be characterized by the resistance of the corresponding rooted tree.

Thus, according to Theorem 3.5.1, the following result holds.

Theorem 3.5.4 *The convergence rate of the inhomogeneous log-linear learning with the optimal schedule presented in Theorem 3.4.2 is estimated as follows:*

$$\|\pi_m - \pi^*\|_{l_1} = O\left(\frac{1}{m^b}\right) \quad m \to \infty,$$

where $b < \min\{1, b'\}$.

The theorem above shows that the convergence rate of the inhomogeneous IndLLL depends on the number of players in the game, their action sets, and the properties of the potential function. If the game's dimension is large or the parameter b', which is dependent on the potential function, is small, the convergence rate may be insufficiently slow. However, one can follow similar argument as in the proof of Theorem 3.5.3 and set the exploration parameter $\epsilon(t)$ in such a way that the rationality parameter $\beta(t)$, recall that $\beta = -\ln\epsilon$, increases as a polynomial function of t. For the independent log-linear learning the following theorem can be formulated.

Theorem 3.5.5 *Let* $\Gamma = (N, \{A_i\}_i, \{U_i\}_i, \phi)$ *be a* discrete action *potential game. Suppose that each player uses the inhomogeneous independent log-linear learning algorithm with the exploration parameter* $\epsilon(S, t) = \frac{S}{\exp\{(t+1)^\alpha\}}$ *for some fixed* α*. Then for any given* $\varepsilon \in (0, 1)$ *there exists* T*, for which* $S = S(T)$ *can be found such that for any* $t \geq T$

$$\|\pi_t - \pi^*\|_{l_1} \leq \varepsilon,$$

where $\pi^* = \lim_{t\to\infty} \pi(t)$ *and* $\pi(t)$ *is the stationary distribution of the time-homogeneous Markov chain* $P(\epsilon(t))$ *of the IndLLL with fixed* t*.*

The simulation results presented in the following subsection support the theoretic analysis in Theorems 3.5.3 and 3.5.5 and demonstrate the effectiveness of the polynomial choice for the exploration parameter $\beta(t)$ in both ILLL and IIndLLL.

3.5.3 Simulation Results: Example of a Sensor Coverage Problem

In this section, the ILLL and IIndLLL algorithms are applied to an example of a modeled sensor coverage problem. The objective is to cover a mission space in the

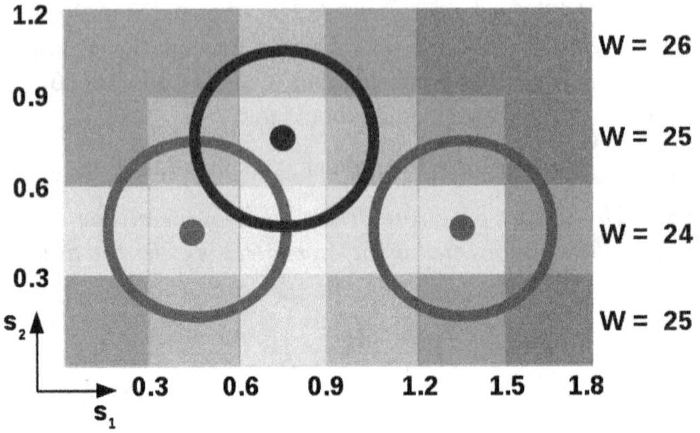

Fig. 3.1 Three agents on the mission space $\mathcal{S}$

most efficient way. Further the sensor coverage problem as it was done in [ZM13b] is formulated, but with some modification.

The mission space $\mathcal{S} \in \mathbb{R}^2$ is the rectangle $\mathcal{S} = [0, 1.8] \times [0, 1.2]$ divided into 6×4 squares. The size of each side is 0.3 (see Fig. 3.1). The rectangle $\mathcal{S}$, thus, can be described by the centers of these squares as follows:

$$\mathcal{S} = \{(0.15 + 0.3k, 0.15 + 0.3l) | k \in \{0, \dots, 5\}, l \in \{0, \dots, 3\}\}.$$

On this mission space, the density function $W(\overline{s}) = e^{\frac{-w\|\overline{s}-\overline{\mu}\|^2}{9}}$, where $\overline{s} = (s_1, s_2) \in \mathcal{S}$, $\overline{\mu} = (0.75; 0.45)$, and

$$w = \begin{cases} 26, & \text{if } s_2 = 0.15, \\ 25, & \text{if } s_2 \in \{0.45, 1.05\}, \\ 24 & \text{if } s_2 = 0.75 \end{cases}$$

is defined. Three agents, which are mobile robots with the sensing radius $r = 0.3$, are placed on the mission space (see Fig. 3.1). The position of each agent is regarded as the action $\overline{a}^i$ to be chosen. Hence, the action set A_i for any $i \in \{1, 2, 3\}$ coincides with $\mathcal{S}$, namely $A_i = \mathcal{S}$. Each agent $i \in \{1, 2, 3\}$ can sense an event at $\overline{s} \in \mathcal{S}$ from her position $\overline{a}^i \in \mathcal{S}$, if $\overline{s} \in D(\overline{a}^i) := \{\overline{s} \in \mathcal{S} : \|\overline{s} - \overline{a}^i\| \leq r\}$. In the example under consideration, the agents can be dislocated in the same position. For each joint action $\boldsymbol{a} = (\overline{a}^1, \overline{a}^2, \overline{a}^3)$ and each $\overline{s} \in \mathcal{S}$ we define $n_{\overline{s}}(\boldsymbol{a})$ as the number of agents such that $\overline{s} \in D(\overline{a}^i)$. As in [ZM13b], the objective of the sensor coverage problem under consideration is described by the function

$$\phi(\boldsymbol{a}) = \sum_{\overline{s} \in \mathcal{S}} \sum_{l=1}^{n_{\overline{s}}(\boldsymbol{a})} \frac{W(\overline{s})}{l}.$$

In [ZM13b] it is shown that the function $\phi(\boldsymbol{a})$ is a potential one, if the agents' utility functions are $\{U_i(\boldsymbol{a}) = \sum_{\bar{s}\in D(a^i)} \frac{W(\bar{s})}{n_{\bar{s}}(\boldsymbol{a})}\}$. Hence, we deal with the potential game $\Gamma = (3, \{A_i = \mathcal{S}\}, \{U_i\}, \phi)$. One can check that in the potential game defined above, there exist two Nash equilibria. One of them is $\boldsymbol{a}_1 = ((0.45, 0.45), (1.05, 0.75), (1.35, 0.75))$, the second one is $\boldsymbol{a}_2 = ((0.75, 0.45), (1.05, 1.05), (1.35, 0.45))$. Since $\phi(\boldsymbol{a}_1) = 8.7343$ and $\phi(\boldsymbol{a}_2) = 7.6318$, we conclude that $\boldsymbol{a}_1$ corresponds to the optimal Nash equilibrium, whereas $\boldsymbol{a}_2$ corresponds to the suboptimal one. The global objective in the system is to get to the optimal state $\boldsymbol{a}_1$.

3.5.3.1 Inhomogeneous Log-Linear Learning in Coverage Problem

Firstly, we apply the inhomogeneous log-linear learning to this problem. According to Theorem 3.3.2, we scale the utility functions such that $U_i \in [0, 1]$ for any i and set the parameter as the logarithmic function $\beta(t) = \frac{\log(t+1)}{c}$, where c is the scrambling constant of the logit dynamics in the system. The implementation result is demonstrated in Fig. 3.2. The graph shows the mean value of the potential function over every 50 iterations. Here, the initial system state is described by $\bar{a}^1(0) = (0.75, 0.15), \bar{a}^2(0) = (0.15, 1.35), \bar{a}^3(0) = (0.45, 0.15)$. We can see that the potential function increases as the algorithm runs. Hence, the learning process leads the system toward its optimal state. However, the increase of the potential function is slow after the moment of getting the value of 7. The graphic shows that after 5000 iterations the potential function is about 90% of the optimal value.

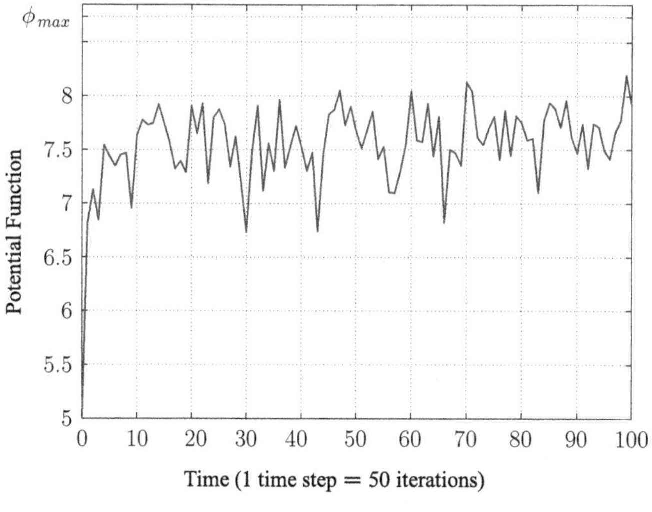

Fig. 3.2 ILLL in coverage problem with logarithmic $\beta(t)$

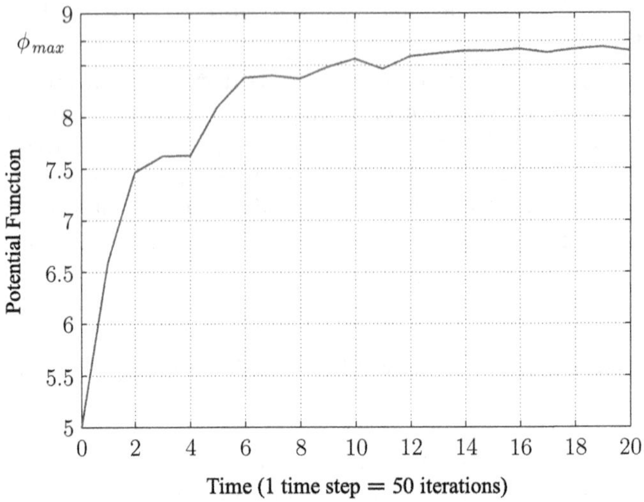

Fig. 3.3 ILLL in coverage problem with polynomial $\beta(t)$

To improve the performance of the algorithm, we turn our attention to Theorem 3.5.3 and change the choice of the parameter β. Figure 3.3 demonstrates the run of the log-linear learning, when $\beta(t) = \frac{t}{80}$ given the same initial state. We can conclude that the potential function is 97% of its optimal value already after 500 iterations. Moreover, it continues to increase during the following 500 iterations.

3.5.3.2 Inhomogeneous Independent Log-Linear Learning in Coverage Problem

To find the optimal dislocation $\boldsymbol{a}_1$ in a distributed *synchronous* manner and not to get stuck in the suboptimal equilibrium $\boldsymbol{a}_2$, agents follow now the rules of the inhomogeneous independent log-linear learning procedure with a time-dependent exploration parameter $\epsilon(t)$.

The initial actions of the agents in our setting are

$$\overline{a}^1(0) = (0.15, 0.15), \overline{a}^2(0) = (0.15, 0.45), \overline{a}^3(0) = (0.45, 0.15).$$

According to Theorem 3.4.2, to guarantee convergence of the algorithm to the optimal state in total variation, one has to set up $\epsilon(t) = (t+1)^{-\frac{3}{10}}$ and $\beta(t) = -\frac{\ln \epsilon(t)}{9}$, given that the utility functions are appropriately scaled. Figure 3.4 demonstrates the result of the IndLLL algorithm implementation with this settings. As before, the mean value of the potential function over every 50 iterations is shown in the graph. Analogously to the case of the classical log-linear learning, we can see that the potential function grows very slowly under this choice of the parameters.

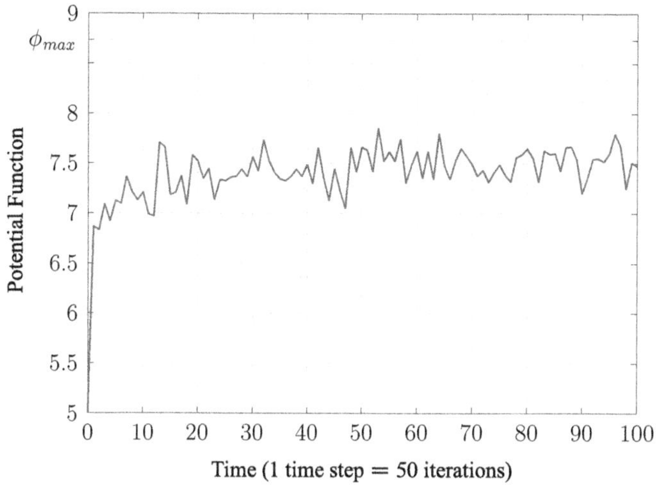

Fig. 3.4 IIndLLL in coverage problem with logarithmic $\beta(t)$

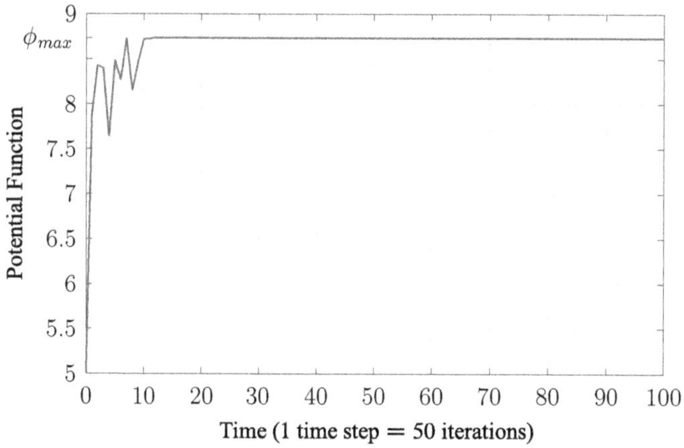

Fig. 3.5 IIndLLL in coverage problem with polynomial $\beta(t)$

That is why the result of Theorem 3.5.5 can be used to set the exploration parameter such that the algorithm demonstrates faster convergence to some state close to the optimal one. Figure 3.5 shows the implementation result of the IndLLL algorithm with $\epsilon(t) = e^{-\frac{t+1}{6}}$ and $\beta(t) = \frac{t+1}{6}$. We can see that in this case the agents get to the optimal state already after 500 iterations and they stay there during the following time steps.

3.6 Learning in Continuous Action Games

This section deals with memoryless learning in potential games with continuous actions. More specifically, the logit dynamics analyzed in the previous section will be extended to continuous actions. Analogously to the case of discrete actions, the memoryless procedures need to be studied by means of Markov chains. The reader with background in continuous state Markov chains can proceed to the discussion in the current section. Otherwise, Appendix A.3 covers the basics of these Markov processes. As before, the focus is on a potential game. Recall that, according to Remark 3.1.1, it can be assumed without loss of generality that utility functions in this game take values in $[-1, 0]$.

3.6.1 Log-Linear Learning in Continuous Action Games

In this subsection the log-linear learning for *continuous action games* is defined. The discrete version of this learning procedure was studied in Sect. 3.3.1. Here a potential game $(N, \{A_i\}, \{U_i\}, \phi)$ is considered, where $[N] = \{1, \ldots, N\}$ is the set of players, whose action sets are represented by finite intervals[4] $A_i = [b_i, c_i] \subset \mathbb{R}$, $i \in [N]$, and utility functions $U_i : \boldsymbol{A} \to [-1, 0]$ are *continuous functions on* $\boldsymbol{A}$ for all $i \in [N]$.

In such game the log-linear learning runs as follows. At each moment of time t one player i is chosen uniformly at random to update her action. The updated action $a^i(t)$ belongs to the subset $B_i \subseteq A_i$ with probability

$$\Pr\{a^i(t) \in B_i\} = \frac{\int_{B_i} \exp\{\beta U_i(x^i, a^{-i}(t-1))\} dx^i}{\int_{A_i} \exp\{\beta U_i(y^i, a^{-i}(t-1))\} dy^i}. \tag{3.26}$$

The rule defined above renders the algorithm a Markov chain P_β with the continuous state space equal to the set of all joint actions $\boldsymbol{A}$. According to this rule, the Markov chain kernel $P_\beta(\boldsymbol{a}, B)$, where $\boldsymbol{a} \in \boldsymbol{A}$ and $B \in \sigma(\boldsymbol{A})$, can be defined as follows:

$$P_\beta(\boldsymbol{a}, B) = \frac{1}{N} \sum_{i=1}^{N} \Pr\{a^i(t) \in B_i\}. \tag{3.27}$$

Here $B_i = l_i(\boldsymbol{a}) \cap B$ is the intersection of the directing line of the i-th coordinate going through $\boldsymbol{a}$ with the set B, $\boldsymbol{a} \in \boldsymbol{A}$, and $\Pr\{a^i(t) \in B_i\}$ is defined as in (3.26) with $a^{-i}(t-1) = a^{-i} = (a^1, \ldots, a^{i-1}, a^{i+1}, \ldots, a^N)$ (see Fig. 3.6).

[4]The following analysis transfers easily to other compact subset of higher dimensions.

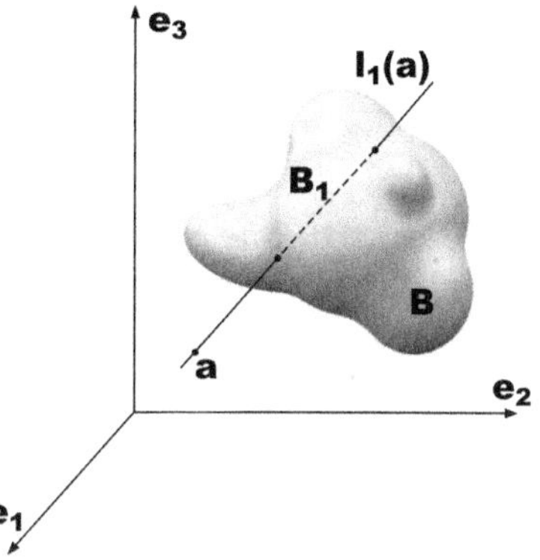

Fig. 3.6 Directing line of 1st coordinate

In the following discussion of this subsection the convergence properties of the Markov chain above will be studied and, thus, behavior of continuous action potential games in which players adhere the rules of the continuous log-linear learning (3.26) will be investigated.

3.6.1.1 Constant Parameter β: Stochastic Stability of Potential Function Maximizers

Firstly, let us focus on time-independent setting of the parameter β in (3.26), (3.27). From the equality (3.27) we conclude that the transition kernel P_β has the following density $p_\beta(\boldsymbol{a},\boldsymbol{x})$:

$$p_\beta(\boldsymbol{a},\boldsymbol{x}) = \begin{cases} \frac{1}{N}\frac{\exp\{\beta U_i(x^i,a^{-i})\}}{\int_{A_i}\exp\{\beta U_i(y^i,a^{-i})\}dy^i} & \text{if } d(\boldsymbol{a},\boldsymbol{x}) = 1 \text{ and } a^i \neq x^i, \\ 0 & \text{otherwise,} \end{cases} \tag{3.28}$$

where $d(\boldsymbol{a},\boldsymbol{x})$ is the Hamming distance between the vectors $\boldsymbol{a}$ and $\boldsymbol{x}$. Thus, in the case of the time-independent parameter β the Markov chain under consideration is homogeneous. Moreover, it is straightforward to check that the probability distribution with the following density:

$$\pi_\beta(\boldsymbol{x}) = \frac{\exp\{\beta\phi(\boldsymbol{x})\}}{\int_A \exp\{\beta\phi(\boldsymbol{y})\}d\boldsymbol{y}} \tag{3.29}$$

corresponds to a stationary measure of the Markov chain P_β. Indeed, let us notice that since the game Γ is potential, $\phi(\boldsymbol{a}) - \phi(\boldsymbol{x}) = U_i(x^i,a^{-i}) - U_i(a^i,x^{-i})$ for any $\boldsymbol{a} = (x^i,a^{-i})$ and $\boldsymbol{x} = (a^i,x^{-i})$, $i \in [N]$, if $a^{-i} = x^{-i}$. Then

$$\pi_\beta(\boldsymbol{a})\frac{\exp\{\beta U_i(x^i,a^{-i})\}}{\int_{A_i}\exp\{\beta U_i(y^i,a^{-i})\}dy^i} = \pi_\beta(\boldsymbol{x})\frac{\exp\{\beta U_i(a^i,x^{-i})\}}{\int_{A_i}\exp\{\beta U_i(y^i,x^{-i})\}dy^i}.$$

Taking into account (3.28), we, thus, get the fulfillment of the detailed balance equation (see (A.2) in Appendix A.3)

$$\pi_\beta(\boldsymbol{a})p_\beta(\boldsymbol{a},\boldsymbol{x}) = \pi_\beta(\boldsymbol{x})p_\beta(\boldsymbol{x},\boldsymbol{a})$$

for the kernel density of the process P_β and the distribution Π_β with the density π_β from (3.29). It allows us to conclude that the distribution Π_β is stationary for the Markov chain P_β given a fixed parameter β (see Theorem A.3.3 in Appendix A.3.1).

Analogously to the Boltzmann stationary distribution in the discrete version of the log-linear learning, the continuous Boltzmann distribution Π_β (also called *Gibbs distribution*) concentrates on global maxima of the potential function ϕ, if ϕ is continuous on the compact set $\boldsymbol{A}$. Formally, in this case (3.29) implies that[5]

$$\Pi_\beta \Rightarrow \Pi^* \text{ as } \beta \to \infty, \text{ where } \Pi^*(\boldsymbol{A}^*) = 1. \tag{3.30}$$

If the set of potential function maximizers $\boldsymbol{A}^*$ has a positive Lebesgue measure $\mathfrak{L}(\boldsymbol{A}^*) > 0$, then the distribution Π^* has the following density:

$$\pi^*(\boldsymbol{x}) = \frac{1}{\mathfrak{L}(\boldsymbol{A}^*)}. \tag{3.31}$$

In other words, Π^* is the uniform distribution on the set $\boldsymbol{A}^*$. Moreover, in this case

$$\lim_{\beta\to\infty} \|\Pi_\beta - \Pi^*\|_{\mathrm{TV}} = 0. \tag{3.32}$$

This follows directly from (3.29) and definition of convergence in total variation (see Appendix A.1).

In the case of finite number m of potential function maximizers, $\boldsymbol{A}^* = \{\boldsymbol{a}_1^*, \ldots, \boldsymbol{a}_m^*\}$, such that all these points are interior elements of $\boldsymbol{A}$ and the Hessian of the function ϕ in these points is negative definite, namely $H_\phi(\boldsymbol{a}_i^*) < 0$, $i \in [m]$, one can use the Laplace method [Hwa80] for the asymptotic estimation of the integral

$$\frac{\int_{\mathcal{O}(\boldsymbol{a}_i^*)} \exp\{\beta\phi(\boldsymbol{x})\}d\boldsymbol{x}}{\int_{\boldsymbol{A}} \exp\{\beta\phi(\boldsymbol{y})\}d\boldsymbol{y}},$$

where $\mathcal{O}(\boldsymbol{a}_i^*)$ is a neighborhood of $\boldsymbol{a}_i^*$ not containing any other maxima $\boldsymbol{a}_j^* \neq \boldsymbol{a}_i^*$. The Laplace method provides the following expression for the limit distribution Π^* for any $i \in [m]$:

$$\Pi^*(\boldsymbol{a}_i^*) = \frac{(\det H_\phi(\boldsymbol{a}_i^*))^{-1/2}}{\sum_{j=1}^m (\det H_\phi(\boldsymbol{a}_j^*))^{-1/2}}. \tag{3.33}$$

For more details on the proof of the formula above the reader is referred to [Hwa80].

[5]The existence of the weak limit here follows from the fact that, as ϕ is continuous on $\boldsymbol{A}$, the sequence $\{\Pi_\beta\}_\beta$ is tight and, thus, Prokhorov's theorem [Pro56] can be applied.

However, to prove convergence of the chain to the distribution Π_β in total variation, one needs to check the condition in Theorem A.3.4, Appendix A.3.1. The following theorem claims that the Markov chain P_β is ergodic and converges to its stationary distribution Π_β.

Theorem 3.6.1 *Consider any* continuous *action potential game* $(N, \{A_i\}, \{U_i\}, \phi)$ *with continuous utility functions* $\{U_i\}$. *The Markov chain* P_β *on the space* $\boldsymbol{A} = A_1 \times \cdots \times A_N$ *defined by* (3.27) *converges in total variation to its unique stationary distribution* Π_β, *whose density is defined by* (3.29), *namely*

$$\lim_{t\to\infty} \|P^t_\beta(\boldsymbol{a}, \cdot) - \Pi_\beta(\cdot)\|_{\mathrm{TV}} = 0$$

for all $\boldsymbol{a} \in \boldsymbol{A}$.

Proof Let $\mathfrak{L}$ be the standard Lebesgue measure on $\boldsymbol{A}$. According to the definition of the Markov chain P_β (3.27), for any $\boldsymbol{a} \in \boldsymbol{A}$ and $B \in \sigma(\boldsymbol{A})$ there exists a finite path

$$\boldsymbol{a} \to B_1 \to \ldots \to B_k \to B,$$

such that

$$P^{k+1}_\beta(\boldsymbol{a}, B) \geq P_\beta(\boldsymbol{a}, B_1)P_\beta(\boldsymbol{a}_1, B_2) \times \cdots \times P_\beta(\boldsymbol{a}_{k-1}, B_k)P_\beta(\boldsymbol{a}_k, B) > 0,$$

where $B_j \in \sigma(\boldsymbol{A}) : \mathfrak{L}(B_j) > 0$, $\boldsymbol{a}_j \in B_j$ for all $j \in [k]$. Thus, according to Definition A.3.1, the chain P_β is $\mathfrak{L}$-irreducible and, hence, the stationary distribution Π_β is unique (see Theorem A.3.1 in Appendix A.3.1). Now we show that P_β is aperiodic. Let us suppose that B_1 and B_2 are disjoint non-empty elements of $\sigma(\boldsymbol{A})$ such that $\Pi_\beta(B_1) > 0$ and $\Pi_\beta(B_2) > 0$, with $\Pi_\beta(\boldsymbol{a}, B_2) = 1$ for all $\boldsymbol{a} \in B_1$. But it would imply, see again (3.27), that $B_2 = \boldsymbol{A}$, since the functions $\{U_i\}$ are continuous on $\boldsymbol{A}$ and, hence, $P_\beta(\boldsymbol{a}, B) < 1$ for any $B \in \sigma(\boldsymbol{A})$: $B \neq \boldsymbol{A}$. The Harris recurrence follows directly from Definition A.3.2 and the fact that, according to continuity of ϕ, $\Pi_\beta(B) = 0$, $B \in \sigma(\boldsymbol{A})$, if and only if $B = \emptyset$. Thus, Theorem A.3.4 implies the result. □

Theorem 3.6.1 claims that, given some fixed parameter β, the continuous log-linear learning converges in total variation to its unique stationary distribution Π_β. Moreover, the limit of Π_β as $\beta \to \infty$ (see (3.30)) guarantees that the algorithm chooses a potential function maximizer with some given in advance probability $p \in (0, 1)$ as time runs, given a constant parameter β that is large enough. However, the theorem above does not imply the algorithm's convergence in some sense to a potential function maximizer with time, since for any fixed value of β the stationary distribution Π_β has the absolutely continuous strictly positive density (3.28) on $\boldsymbol{A}$ and, hence, $\Pi_\beta(B) > 0$ for any $B \in \sigma(\boldsymbol{A}) : B \neq \emptyset$. Thus, analogously to the discrete version of the log-linear learning, one needs to set up a time-dependent parameter $\beta(t)$ to investigate the algorithm's convergence toward an optimal Nash equilibrium in the potential game.

3.6.1.2 Time-Dependent Parameter β: Convergence to Potential Function Maximizers

Now we proceed to set up the parameter $\beta(t)$ in the continuous log-linear learning as a function of time. Under such choice of the parameter, the corresponding Markov chain is time-inhomogeneous. Thus, we would like to choose the algorithm parameter $\beta(t)$ in such a way that the state of the Markov chain steers toward an optimal system's state expressed by a potential function maximizer. To simplify notations, we denote further the one-step kernel at the time step t of the Markov chain by $P_{\beta(t)}$, and the kernel of the k transitions after the moment of time t by $P_\beta^{t,k}$.

1. **Weak ergodicity.** First of all, we show that there exists such setting of the parameter $\beta(t)$ that the Markov chains defined by (3.27) are weakly ergodic. The following lemma formulates this claim.

Theorem 3.6.2 *Let* $\Gamma = (N, \{A_i\}_i, \{U_i\}_i, \phi)$ *be a* continuous *action potential game with the utility functions* $U_i : \boldsymbol{A} \to [-1, 0]$. *Then there exists such a constant* c *that the log-linear learning algorithm with the parameter* $\beta(t) = \frac{\ln(t+1)}{c}$ *defines the Markov chain that is weakly ergodic.*

Proof According to the definition of the continuous log-linear learning (3.26), (3.27) a time-dependent parameter does not change the pattern of Markov chain in the following sense. For any $t \geq 1$, $\boldsymbol{a} \in \boldsymbol{A}$, and $B \in \sigma(\boldsymbol{A})$, the probability $P_{\beta(t)}(\boldsymbol{a}, B)$ is strictly positive if and only if so is the probability $P_{\beta(t-1)}(\boldsymbol{a}, B)$. For any fixed parameter β the log-linear learning procedure corresponds to the ergodic time-homogeneous Markov chain (Theorem 3.6.1), whose pattern does not depend on β, i.e., $P_{\beta_1}(\boldsymbol{a}, B) > 0$, $\beta_1 \geq 0$, $\boldsymbol{a} \in \boldsymbol{A}$, and $B \in \sigma(\boldsymbol{A})$, if and only if $P_{\beta_2}(\boldsymbol{a}, B) > 0$, for any $\beta_2 \geq 0$. Hence, there exists a positive constant integer c such that the coefficient of ergodicity $\tau(P_\beta^{t,c})$ is strictly less than 1 (see Lemma A.3.4 in Appendix A.3.3) for any parameter β (that can be also time-dependent), and any initial moment of time t. Moreover, according to Definition A.3 in Appendix A.3.3, the following is fulfilled. There exists the state $\boldsymbol{a}' \in \boldsymbol{A}$ and $B' \in \sigma(\boldsymbol{A})$ such that

$$\alpha(P_\beta^{t,c}) = 1 - \tau(P_\beta^{t,c}) \geq P_\beta^{t,c}(\boldsymbol{a}', B') > 0. \tag{3.34}$$

Next we demonstrate that the time-inhomogeneous chain $P_{\beta(t)}$ defined by the log-linear learning with the parameter $\beta(t) = \frac{\ln(t+1)}{c}$, where c is defined above, is weakly ergodic. Indeed,

$$P_\beta^{t,c}(\boldsymbol{a}', B') = \int_{\boldsymbol{A}} [P_{\beta(t)}(\boldsymbol{a}', d\boldsymbol{y}_1) \int_{\boldsymbol{A}} [P_{\beta(t+1)}(\boldsymbol{y}_1, d\boldsymbol{y}_2) \times \cdots \int_{B'} P_{\beta(t+c-1)}(\boldsymbol{y}_{c-1}, d\boldsymbol{y})] \ldots] \tag{3.35}$$

where, according to the choice of $\beta(t)$, for any time step t, $\boldsymbol{a} \in \boldsymbol{A}$ and $C \in \sigma(\boldsymbol{A})$

$$\int_C P_{\beta(t)}(\boldsymbol{a}, d\boldsymbol{y}) = \frac{1}{N} \sum_{i=1}^{N} \frac{\int_{l_i(\boldsymbol{a}) \cap C} (t+1)^{\frac{U_i(y^i, \boldsymbol{a}^{-i})}{c}} dy^i}{\int_{A_i} (t+1)^{\frac{U_i(y^i, \boldsymbol{a}^{-i})}{c}} dy^i}.$$

Since $U_i(\boldsymbol{a}) \in [-1, 0]$ and $P_\beta^{t,c}(\boldsymbol{a}', B') > 0$, the integral $\int_{B'} P_{\beta(t+c-1)}(\boldsymbol{y}_{c-1}, d\boldsymbol{y})$ can be bounded as follows:

$$\int_{B'} P_{\beta(t+c-1)}(\boldsymbol{y}_{c-1}, d\boldsymbol{y}) \geq \frac{1}{N} \frac{\mu_L(l_i(\boldsymbol{y}_{c-1}) \cap B')}{\mu_L(A_i)(t+c)^{\frac{1}{c}}},$$

for some i such that $\mu_L(l_i(\boldsymbol{y}_{c-1}) \cap B') > 0$. We can continue to bound each integral in (3.35) in the same way. Finally, we get

$$P_\beta^{t,c}(\boldsymbol{a}', B') \geq K(N, c, B^*) \prod_{j=t}^{t+c-1} \frac{1}{(j+1)^{\frac{1}{c}}} \geq K(N, c, B^*) \frac{1}{t+c},$$

where $K(N, c, B^*)$ is a constant dependent on N, c, and B^*. Hence, taking (3.34) into account, we get

$$\sum_{j=0}^{\infty} \alpha(P_\beta^{jc,c}) > \sum_{j=0}^{\infty} K(N, c, B^*) \frac{1}{jc+c} = \infty.$$

Thus, the criterion formulated in Theorem A.3.6 implies the weak ergodicity of $P_{\beta(t)}$ with $\beta(t) = \frac{\ln(t+1)}{c}$. □

According to Theorems 3.3.2 and 3.6.2, there exists some constant c that the log-linear learning algorithm with the time-dependent parameter $\beta(t) = \frac{\ln(t+1)}{c}$ initiates a weakly ergodic Markov chain in both discrete and continuous action games. The constant c depends on the properties of a concrete multi-agent system and respectively a modeled potential game. In the case of finite discrete actions (see Theorem 3.3.2), c corresponds to the scrambling constant of the Markov chain. This scrambling constant can be determined as follows. One constructs the "pattern matrix" $\mathfrak{P}(\Gamma)$ of the corresponding transition probability matrix in the potential game Γ under consideration. This "pattern matrix" is the transition probability matrix of the log-linear learning in Γ with $\beta(t) = 0$. The constant c, thus, can be defined as a minimal integer c such that $\alpha(\mathfrak{P}^c(\Gamma)) > 0$ where $\alpha(\cdot) = 1 - \tau(\cdot)$ and $\tau(\cdot)$ is the coefficient of ergodicity of the discrete state Markov chain. Unfortunately, it is impossible to repeat the same procedure for the Markov chain P_β initiated by the log-linear learning in continuous action games. Here, we present a universal method for the constant c determination.

Let us turn back again to the Markov chain P_β defined by (3.26), (3.27). As we have mentioned in the proof of Theorem 3.6.2, the pattern of P_β is the same for all parameters β. Hence, $\alpha(P^{t,c}_{\beta}) > 0$ for any β, if and only if $\alpha(P^{0,c}_{\beta=0}) > 0$. That is why we consider here only the log-linear learning with $\beta = 0$ and search for the possibly smallest constant c such that $\alpha(P^{0,c}_{\beta=0}) > 0$.

Further the result formulated in [RR04] is used.

Theorem 3.6.3 *Consider a Markov chain defined on the Lebesgue measurable space $(\mathcal{X}, \sigma(\mathcal{X}), \mu_L)$ having the unique stationary distribution $\Pi(\cdot)$. Suppose that for some $n_0 \in \mathbb{N}$, $\epsilon > 0$, and probability measure $\nu(\cdot)$ the following condition holds:*

$$P^{n_0}(x, B) \geq \epsilon\nu(B), \quad \text{for all } \boldsymbol{x} \in \mathcal{X}, B \in \sigma(\mathcal{X}). \tag{3.36}$$

Then for all $\boldsymbol{x} \in \mathcal{X}$

$$\|P^c(\boldsymbol{x}, \cdot) - \Pi(\cdot)\|_{\mathrm{TV}} \leq (1-\epsilon)^{\lfloor c/n_0 \rfloor}, \quad c = 1, 2, \ldots.$$ [6]

According to one of the equivalent definitions of the coefficient of ergodicity (A.4) (see Appendix A.3.3) and the triangle inequality,

$$1 - \alpha(P^c_{\beta=0}) \leq 2 \sup_{\boldsymbol{a} \in A} \|P^{0,c}_{\beta=0}(\boldsymbol{a}, \cdot) - \Pi_{\beta=0}(\cdot)\|_{\mathrm{TV}}.$$

Thus, to define an integer $c > 0$ such that $\alpha(P^{0,c}_{\beta=0}) > 0$ one needs to find c such that $\sup \|P^{0,c}_{\beta=0}(\boldsymbol{a}, \cdot) - \Pi_{\beta=0}(\cdot)\|_{\mathrm{TV}} < \frac{1}{2}$. Theorem 3.6.3 implies

$$\|P^c_{\beta=0}(\boldsymbol{a}, \cdot) - \Pi_{\beta=0}(\cdot)\|_{\mathrm{TV}} \leq (1-\epsilon)^{\lfloor c/n_0 \rfloor}.$$

Hence, once n_0 and ϵ are known, the integer c^* can be determined by the inequality $(1-\epsilon)^{\lfloor c/n_0 \rfloor} \leq \frac{1}{2+\varepsilon}$ for some $\varepsilon > 0$. For such c^* coefficient of ergodicity $\tau(P^{0,c^*}_{\beta=0})$ is strictly less than 1.

Now it will be shown that for the log-linear learning applied to a continuous action game with N players, the inequality (3.36) of Theorem 3.6.3 holds for some ϵ and probabilistic measure $\nu(\cdot)$, if we put $n_0 = N$. As $\beta = 0$, $P_{\beta=0}(\boldsymbol{a}, B) = \frac{1}{N}\sum_{i=1}^{N} \frac{\mu_L(l_i(\boldsymbol{a}) \cap B)}{\mu_L(A_i)}$, hence,

$$\begin{aligned} &P^N(\boldsymbol{a}, B) \\ &= \int_A \left[P_{\beta=0}(\boldsymbol{a}, d\boldsymbol{y}_1) \times \cdots \times \int_A \left[P_{\beta=0}(\boldsymbol{y}_{N-2}, d\boldsymbol{y}_{N-1}) \left(\frac{1}{N} \sum_{i=1}^{N} \frac{\mu_L(l_i(\boldsymbol{y}_{N-1}) \cap B)}{\mu_L(A_i)} \right) \right] \cdots \right] \\ &\geq \frac{1}{N^N} \frac{\mu_L(B)}{\mu_L(A_1) \ldots \mu_L(A_N)} = \frac{1}{N^N} \frac{\mu_L(B)}{\mu_L(\boldsymbol{A})}. \end{aligned}$$

[6] $\lfloor r \rfloor$ is the greatest integer not exceeding r.

Putting in Theorem 3.6.3 $n_0 = N$, $\nu(\cdot) = \frac{\mu_L(\cdot)}{\mu_L(\boldsymbol{A})}$, and $\epsilon = \frac{1}{N^N}$, we obtain the following result:

Proposition 3.6.1 *The integer c in Theorem 3.6.2 can be chosen as the minimal integer c such that* $\left(1 - \frac{1}{N^N}\right)^{\lfloor c/N \rfloor} \leq \frac{1}{2+\varepsilon}$ *is fulfilled for any* $\varepsilon > 0$.

The proposition above provides the universal way for the constant c determination, figuring in the function $\beta(t) = \frac{\ln(t+1)}{c}$. Recall that this setting of the parameter in the log-linear learning guarantees the weak ergodicity of the corresponding Markov chain.

It has been demonstrated that the setting for the algorithm's parameter $\beta(t)$ exists such that the Markov chain $P_{\beta(t)}$ is weakly ergodic. However, the weak ergodicity means only that the information on any initial state is forgotten with time (see Appendix A.3.3), but does not imply convergence of Markov chain $P_{\beta(t)}$ to some state independent of its initial one. In the following, the work investigates the question of the Markov chain convergence and, hence, the log-linear algorithm convergence, to the set of potential function maximizers in a continuous action potential game regardless of the initial joint action.

2. **Positive Lebesgue measure of the set A^* Convergence in total variation of the log-linear learning.** First, we study the case when the set of potential function maximizers has positive Lebesgue measure, i.e., $\mathfrak{L}(\boldsymbol{A}^*) > 0$. The main result in this case can be formulated by the following theorem.

Theorem 3.6.4 *Let* $\Gamma = (N, \{A_i\}_i, \{U_i\}_i, \phi)$ *be a* continuous *action potential game, where* $\boldsymbol{A}$ *is compact, the utility functions* $U_i : \boldsymbol{A} \to [-1, 0]$ *are continuous on* $\boldsymbol{A}$, *the set* $\boldsymbol{A}^* = \{\boldsymbol{a}^* \in \boldsymbol{A} : \boldsymbol{a}^* = \arg\max_{\boldsymbol{A}} \phi(\boldsymbol{a})\}$ *has positive Lebesgue measure, and* $\max_{\boldsymbol{A}} \phi(\boldsymbol{a}) = 0$. *Then the Markov chain* P_β *initiated by the continuous log-linear learning* (3.26) *is strongly ergodic, given* $\beta(t) = \frac{\ln(t+1)}{c}$, *where c is defined as in Proposition 3.6.1. Moreover, the unique stationary measure* Π^*_β *of this chain is uniformly distributed on* $\boldsymbol{A}^*$.

Proof To prove this claim, we will use the result formulated in Theorem A.3.7, see Appendix A.3.3. According to Theorems 3.6.1 and 3.6.2, conditions (1) and (2) of Theorem A.3.7 hold.

Next we show the fulfillment of condition (A.3.7) of Theorem A.3.7. Let for any t the measure Π_t denote the distribution with the density $\pi_{\beta(t)}$ defined in (3.5), when $\beta(t) = \frac{\ln(t+1)}{c}$. Then, according to the formula for the total variation distance (see Theorem A.1.1, Appendix A.1), we get

$$\|\Pi_t(\cdot) - \Pi_{t-1}(\cdot)\|_{\mathrm{TV}} = \int_A |\pi_{\beta(t)}(\boldsymbol{x}) - \pi_{\beta(t-1)}(\boldsymbol{x})| d\boldsymbol{x},$$

where

$$\pi_{\beta(t-1)}(\boldsymbol{x}) = \frac{t^{\frac{\phi(\boldsymbol{x})}{c}}}{\int_A t^{\frac{\phi(\boldsymbol{y})}{c}} d\boldsymbol{y}} = \frac{1}{\int_A t^{\frac{\phi(\boldsymbol{y}) - \phi(\boldsymbol{x})}{c}} d\boldsymbol{y}}.$$

Let us consider the integral in the denominator of the above fraction, namely

$$I_x(t) = \int_A t^{\frac{\phi(y)-\phi(x)}{c}} dy.$$

Since the set A is compact, the integrand here is a continuous function of (t, y) and so is its partial derivative. Hence, $I_x(t)$ can be differentiated with respect to the parameter t under the integral sign as follows:

$$\frac{\partial}{\partial t} I_x(t) = \int_A \frac{\phi(y) - \phi(x)}{c} t^{\frac{\phi(y)-\phi(x)}{c} - 1} dy.$$

Note that, if $x \in A^*$, then $\frac{\partial}{\partial t} I_x(t) \leq 0$ and, hence, $I_x(t)$ decreases with respect to t. Otherwise, i.e., if $x \notin A^*$, there exists $\hat{A} \subseteq A$ with $\mathfrak{L}(\hat{A}) \neq 0$ such that $\phi(y) \geq \phi(x)$, if and only if $y \in \hat{A}$. Let $M(x) = \max_{y \in A \setminus \hat{A}} |\phi(y) - \phi(x)|$. Then

$$\frac{\partial}{\partial t} I_x(t) \geq \left(\int_{\hat{A}} \frac{\phi(y) - \phi(x)}{c} t^{\frac{\phi(y)-\phi(x)}{c}} dy \right) t^{-1} - \frac{M(x)\mu(A \setminus \hat{A})}{c} t^{-\frac{M(x)}{c} - 1},$$

that is strictly positive starting from some t_0.

Thus, there exists such t_0 that for any $t \geq t_0$ the following is true. If $x \in A^*$, then $I_x(t)$ decreases. If $x \in A \setminus A^*$, then $I_x(t)$ increases. Hence, taking into account $\mathfrak{L}(A^*) > 0$, we can write

$$\begin{aligned}
\sum_{t=1}^{\infty} \int_A |\pi_{\beta(t)}(x) - \pi_{\beta(t-1)}(x)| dx &= \sum_{t=1}^{t_0} \int_A |\pi_{\beta(t)}(x) - \pi_{\beta(t-1)}(x)| dx \\
&+ \sum_{t=t_0+1}^{\infty} \int_{A^*} (\pi_{\beta(t)}(x) - \pi_{\beta(t-1)}(x)) dx \\
&+ \sum_{t=t_0+1}^{\infty} \int_{A \setminus A^*} (\pi_{\beta(t-1)}(x) - \pi_{\beta(t)}(x)) dx \\
&= S(t_0) + 2 \int_{A^*} \lim_{t \to \infty} \pi_{\beta(t)}(x) dx < \infty,
\end{aligned}$$

where $S(t_0) = \sum_{t=1}^{t_0} \int_A |\pi_t(x) - \pi_{t-1}(x)| dx$. Thus, condition (A.3.7) of Theorem A.3.7 holds. Thus, Theorem A.3.7 together with (3.31) and (3.32) allows us to conclude the strong ergodicity of the Markov chain P_β under the choice $\beta = \beta(t) = \frac{\ln(t+1)}{c}$. □

The theorem above claims that in the case of positive Lebesgue measure of A^*, $\mathfrak{L}(A^*) > 0$, the log-linear learning defined by (3.26) converges in total variation to the uniform distribution over optimal Nash equilibria, namely over the set A^*.

In the following it is shown that this convergence does not take place, if $\mathfrak{L}(A^*) = 0$. Nevertheless, it is demonstrated that, in this case, the continuous log-linear learning converges in the weak sense to a potential function maximizer being applied to a potential game with continuous actions.

3. **Zero Lebesgue measure of the set A^*. Weak convergence of the log-linear learning.** Now let us focus on the case $\mathfrak{L}(\boldsymbol{A}^*) = 0$. Let Π^0 be any measure with full support on the set $\boldsymbol{A}^*$ and the parameter $\beta = \beta(t)$ is chosen to be such time-dependent function that $\lim_{t\to\infty} \beta(t) = \infty$. Since $\mathfrak{L}(\boldsymbol{A}^*) = 0$ and, thus, $P^t_\beta(\boldsymbol{a}, B) = 0$ for any t, $\boldsymbol{a} \in \boldsymbol{A}$, and $B \subseteq \boldsymbol{A}^*$,

$$\lim_{t\to\infty} \|P^t_\beta(\boldsymbol{a}, \cdot) - \Pi^0(\cdot)\|_{\mathrm{TV}} = 1 \quad \text{for any } \boldsymbol{a} \in \boldsymbol{A}.$$

Thus, there is no convergence in total variation of the Markov chain P_β to Π^0 as time runs.

We proceed to investigate the convergence properties of the Markov chain P_β with $\beta = \beta(t) = \frac{\ln(t+1)}{c}$ for the case $\mathfrak{L}(\boldsymbol{A}^*) = 0$. According to Theorem 3.6.2, this Markov chain is weakly ergodic. The next theorem claims that in addition to the weak ergodicity, the Markov chain under consideration converges in total variation to the distribution $\Pi_{\beta(t)}$ with the density $\pi_{\beta(t)}$ defined in (3.29) as time runs and given $\beta(t) = \frac{\ln(t+1)}{c}$. Let $\Pi(t)$ be the state of P_β after t steps given an initial distribution $\Pi(0)$, i.e., $\Pi(t) = \Pi(0)P^{0,t}_\beta$.

Theorem 3.6.5 *Let $\Gamma = (N, \{A_i\}_i, \{U_i\}_i, \phi)$ be a* continuous *action potential game, where $\boldsymbol{A}$ is compact, the utility functions $U_i : \boldsymbol{A} \to [-1, 0]$ are continuous on $\boldsymbol{A}$, the set $\boldsymbol{A}^* = \{\boldsymbol{a}^* \in \boldsymbol{A} : \boldsymbol{a}^* = \arg\max_{\boldsymbol{A}} \phi(\boldsymbol{a})\}$ has zero Lebesgue measure, and $\max_{\boldsymbol{A}} \phi(\boldsymbol{a}) = 0$. Then the Markov chain P_β defined by the continuous version of the log-linear learning with $\beta = \beta(t) = \frac{\ln(t+1)}{c}$ is such that for any initial distribution $\Pi(0)$ on $\boldsymbol{A}$*

$$\lim_{t\to\infty} \|\Pi(t) - \Pi_{\beta(t)}\|_{\mathrm{TV}} = 0.$$

To prove this theorem we need the following lemma:

Lemma 3.6.1 *Let $\Pi_{\beta(t)}$ be the distribution with the density $\pi_{\beta(t)}$ defined in* (3.29). *Then, if ϕ is a continuous function defined on the compact set $\boldsymbol{A}$,*

$$\|\Pi_{\beta(t)} - \Pi_{\beta(t-1)}\|_{\mathrm{TV}} \le 2\log\left(\frac{I(\beta(t))}{I(\beta(t-1))}\right), \tag{3.37}$$

where $I(\beta) = \int_A e^{\beta\phi(\boldsymbol{x})}d\boldsymbol{x}$.

Proof According to the formula for total variation distance (Theorem A.1.1 in Appendix A.1) and (3.29),

$$\|\Pi_{\beta(t)} - \Pi_{\beta(t-1)}\|_{\mathrm{TV}} = \int_{A} D(\boldsymbol{x})d\boldsymbol{x},$$

$$\text{where } D(\boldsymbol{x}) = \left| \frac{e^{\beta(t)\phi(\boldsymbol{x})}}{I(\beta(t))} - \frac{e^{\beta(t-1)\phi(\boldsymbol{x})}}{I(\beta(t-1))} \right|.$$

Since ϕ is continuous on $\boldsymbol{A}$ and $\boldsymbol{A}$ is compact, $I(\beta)$ can be differentiated under the integral sign and $I'(\beta) = \frac{d}{d\beta}I(\beta) = \int_{A} \phi(\boldsymbol{x})e^{\beta\phi(\boldsymbol{x})}d\boldsymbol{x}$. Hence,

$$D(\boldsymbol{x}) = \left| \int_{\beta(t-1)}^{\beta(t)} \frac{d}{dy}\left(\frac{e^{y\phi(\boldsymbol{x})}}{I(y)}\right)dy \right| = \left| \int_{\beta(t-1)}^{\beta(t)} \frac{\phi(\boldsymbol{x})e^{y\phi(\boldsymbol{x})}I(y) - e^{y\phi(\boldsymbol{x})}\int_{A}\phi(\boldsymbol{x})e^{y\phi(\boldsymbol{x})}d\boldsymbol{x}}{I^2(y)}dy \right|$$

$$\leq \int_{\beta(t-1)}^{\beta(t)} \left(\frac{\phi(\boldsymbol{x})e^{y\phi(\boldsymbol{x})}}{I(y)} + e^{y\phi(\boldsymbol{x})}\frac{I'(y)}{I^2(y)} \right) dy$$

Thus, we get

$$\|\Pi_{\beta(t)} - \Pi_{\beta(t-1)}\|_{\mathrm{TV}} = \int_{A} D(\boldsymbol{x})d\boldsymbol{x} \leq \int_{\beta(t-1)}^{\beta(t)} \frac{2I'(y)}{I(y)}dy = 2\log\left(\frac{I(\beta(t))}{I(\beta(t-1))}\right).$$

□

Remark 3.6.1 Note that the lemma above does not use any information about the measure of the set $\boldsymbol{A}^*$ and, thus, proposes another way to prove the statement in Theorem 3.6.4 in the case of the following assumption on the utility and potential functions: $U_i, \phi : \boldsymbol{A} \to [-1, 0]$ for any $i \in [N]$, $\phi(\boldsymbol{a}^*) = 0$ for any $\boldsymbol{a}^* \in \boldsymbol{A}$, and $\mathfrak{L}(\boldsymbol{A}^*) > 0$. Indeed, because of (3.37),

$$\sum_{t=1}^{\infty} \|\Pi_{\beta(t)} - \Pi_{\beta(t-1)}\|_{\mathrm{TV}} = \lim_{t\to\infty} 2\log\left(\frac{I(\beta(t))}{I(\beta(0))}\right).$$

Since $\phi(\boldsymbol{a}^*) = 0$ for any $\boldsymbol{a}^* \in \boldsymbol{A}$, $\mathfrak{L}(\boldsymbol{A}^*) > 0$, and $\phi(\boldsymbol{x}) < 0$, if $\boldsymbol{x} \in \boldsymbol{A} \setminus \boldsymbol{A}^*$,

$$\lim_{\beta(t)\to\infty} I(\beta(t)) = \mathfrak{L}(\boldsymbol{A}^*) + \lim_{\beta(t)\to\infty} \int_{\boldsymbol{A}\setminus\boldsymbol{A}^*} e^{\beta(t)\phi(\boldsymbol{x})}d\boldsymbol{x} = \mathfrak{L}(\boldsymbol{A}^*).$$

Hence,

$$\sum_{t=1}^{\infty} \|\Pi_{\beta(t)} - \Pi_{\beta(t-1)}\|_{\mathrm{TV}} = 2\log\left(\frac{\mathfrak{L}(\boldsymbol{A}^*)}{I(\beta(0))}\right) < \infty.$$

Proof of Theorem 3.6.5. Let the sequence $n_k = ck$, $r_k = [k - k^{1-\delta/2}]$, where $0 < \delta < 1$ and $k \in \mathbb{N}$. In Corollary 5.2 of [HS91][7] it was proven that the sequence $\{|\beta(n_{k+2}) - \beta(n_{r_k})|\}_k$ is bounded given $\beta(t) = \frac{\ln(t+1)}{c}$. On the other hand, the Laplace asymptotic of the integral $\int_A e^{\beta\phi(x)}d\boldsymbol{x}$, the fact that $\phi(a^*) = 0$ for any $\boldsymbol{a}^* \in \boldsymbol{A}^*$, and the boundedness of the sequence $\{|\beta(n_{k+2}) - \beta(n_{r_k})|\}_k$ allow us to conclude that

$$\lim_{k\to\infty} \frac{I(\beta(n_{k+2}))}{I(\beta(n_{r_k}))} = 1$$

and, hence, according to (3.37),

$$\sum_{t=n_{r_k}+1}^{n_{k+2}} \|\Pi_{\beta(t)} - \Pi_{\beta(t-1)}\|_{\mathrm{TV}} \to 0 \text{ as } k \to \infty. \tag{3.38}$$

Let us fix any $\varepsilon > 0$. Since the Markov chain P_β with $\beta = \beta(t) = \frac{\ln(t+1)}{c}$ is weakly ergodic and $k - r_k \to \infty$ as $k \to \infty$, there exists such K that for any $k > K$

$$\tau\left(P_\beta^{n_{r_k}, n_{k+1}}\right) < \frac{\varepsilon}{4} \tag{3.39}$$

and, according to (3.38),

$$\sum_{t=n_{r_k}+1}^{n_{k+2}} \|\Pi_{\beta(t)} - \Pi_{\beta(t-1)}\|_{\mathrm{TV}} < \frac{\varepsilon}{2}. \tag{3.40}$$

Now let $m > n_{K+1}$. Then there exists such $k' > K$ that $n_{k'+1} < m \le n_{k'+2}$. Using the triangle inequality and the argument analogous to one in (3.23) of Theorem 3.5.3, we can write:

$$\begin{aligned}
\|\Pi(m) - \Pi_{\beta(m)}\|_{\mathrm{TV}} &= \|\Pi(n_{r_{k'}}) P_\beta^{n_{r_{k'}}, m} - \Pi_{\beta(m)}\|_{\mathrm{TV}} \\
&\le \|(\Pi(n_{r_{k'}}) - \Pi_{\beta(n_{r_{k'}})}) P_\beta^{n_{r_{k'}}, m}\|_{\mathrm{TV}} + \|\Pi_{\beta(n_{r_{k'}})} P_\beta^{n_{r_{k'}}, m} - \Pi_{\beta(m)}\|_{\mathrm{TV}} \\
&\le 2\tau\left(P_\beta^{n_{r_{k'}}, m}\right) + \sum_{t=n_{r_{k'}}+1}^{m} \|(\Pi_{\beta(t)} - \Pi_{\beta(t-1)}) P_\beta^{t-1, m}\|_{\mathrm{TV}} \\
&\le 2\tau\left(P_\beta^{n_{r_{k'}}, n_{k'+1}}\right) + \sum_{t=n_{r_{k'}}+1}^{n_{k'+2}} \|\Pi_{\beta(t)} - \Pi_{\beta(t-1)}\|_{\mathrm{TV}} < \varepsilon,
\end{aligned}$$

[7] In the work [HS91] the authors study the simulating annealing algorithm over general spaces. The proof of Theorem 3.6.5 follows the same line as one of Theorem 5.1 in [HS91].

where the last inequality is due to (3.39) and (3.40). Since ε is chosen arbitrary small, the result follows. □

Thus, it has been shown that, if $\mathfrak{L}(\boldsymbol{A}^*) = 0$, the Markov chain of the continuous log-linear learning tracks the distribution $\Pi_{\beta(t)}$ as time runs given an appropriate setting of the time-dependent parameter $\beta(t)$. Convergence to this distribution is in total variation. Taking into account that $\Pi_\beta \Rightarrow \Pi^*$, where $\Pi^*(\boldsymbol{A}^*) = 1$, given the continuous function ϕ on the compact $\boldsymbol{A}$ (see (3.30)), the following corollary from Theorem 3.6.5 is obtained.

Corollary 3.6.1 *Let* $\Gamma = (N, \{A_i\}_i, \{U_i\}_i, \phi)$ *be a* continuous *action potential game, where* $\boldsymbol{A}$ *is compact, the utility functions* $U_i : \boldsymbol{A} \to [-1, 0]$ *are continuous on* $\boldsymbol{A}$*, the set* $\boldsymbol{A}^* = \{\boldsymbol{a}^* \in \boldsymbol{A} : \boldsymbol{a}^* = \arg\max_{\boldsymbol{A}} \phi(\boldsymbol{a})\}$ *has zero Lebesgue measure, and* $\max_{\boldsymbol{A}} \phi(\boldsymbol{a}) = 0$*. Then the Markov chain* $P_{\beta(t)}$ *initiated by the continuous log-linear learning* (3.26) *converges weakly to the probability measure* $\Pi^* : \Pi^*(\boldsymbol{A}^*) = 1$*, i.e.,*

$$\Pi(t) \Rightarrow \Pi^* \quad \textit{as } t \to \infty,$$

given $\beta(t) = \frac{\ln(t+1)}{c}$*, where c is defined as in Proposition 3.6.1 and* Π^* *is defined in* (3.33).

The corollary above claims the weak convergence of the log-linear learning to a potential function maximizer, being applied to a potential game with continuous compact action sets and continuous utility functions. Thus, taking into account the Portmanteau Lemma (see Lemma A.1.2 in Appendix A.1), we can conclude that under the log-linear learning the joint action $\boldsymbol{a}(t)$ is such that for any $\varepsilon > 0$

$$\lim_{t\to\infty} \Pr\{\boldsymbol{a}(t) \in \overline{B}^{\varepsilon}_{\boldsymbol{A}^*}\} = 0,$$

where $\overline{B}^{\varepsilon}_{\boldsymbol{A}^*} = \{\boldsymbol{a} \in \boldsymbol{A} : \|\boldsymbol{a} - \boldsymbol{a}^*\| \geq \varepsilon \text{ for all } \boldsymbol{a}^* \in \boldsymbol{A}^*\}$.

3.6.2 Independent Log-Linear Learning in Continuous Action Games

This subsection defines the independent log-linear learning for *continuous action games*, whose discrete version was introduced in Sect. 3.4.2. Throughout this subsection again a potential game $(N, \{A_i\}, \{U_i\}, \phi)$ is considered, where $[N] = \{1, \ldots, N\}$ is the set of the players, whose action sets are represented by finite intervals[8] $A_i = [b_i, c_i] \subset \mathbb{R}$, $i \in [N]$, and utility functions $U_i : \boldsymbol{A} \to [-1, 0]$ are *continuous functions on* $\boldsymbol{A}$ for all $i \in [N]$.

[8] The analysis can be generalized to the case of compact sets of any dimension.

In such game the independent log-linear learning runs as follows. At each moment of time t every player i, $i \in [N]$, independently of each other updates her action $a^i(t)$ according to the probabilistic distribution below:

$$a^i(t) = \begin{cases} a^i(t-1) & \text{w.p. } 1 - \epsilon^m, \\ a^i \in B_i & \text{w.p. } \epsilon^m \frac{\int_{B_i} \exp\{\beta U_i(x, \boldsymbol{a}^{-i}(t-1))\}dx}{\int_{A_i} \exp\{\beta U_i(x, \boldsymbol{a}^{-i}(t-1))\}dx}, \end{cases} \tag{3.41}$$

where $B_i \in \sigma(A_i)$, β and $\epsilon < 1$ are the rationality and exploration parameters correspondingly, and m is some constant. Let us choose β such that $\beta = -\ln \epsilon$. Then the probabilistic distribution (3.41) has the following density:

$$p_{i,\epsilon}(a^i, x) = (1 - \epsilon^m)\delta_{a^i}(x) + \epsilon^m \frac{\epsilon^{-U_i(x, \boldsymbol{a}^{-i})}}{\int_{A_i} \epsilon^{-U_i(x, \boldsymbol{a}^{-i})} dx}, \tag{3.42}$$

where $a^i = a^i(t-1)$. According to (3.41) and (3.42), the process described above is a Markov chain G_ϵ defined on the continuous set of joint actions $\boldsymbol{A} = A_1 \times \cdots \times A_N$. Since agents act independently, the transition probability kernel $G_\epsilon(\boldsymbol{a}, C)$, $\boldsymbol{a} \in \boldsymbol{A}$ and $C \in \sigma(\boldsymbol{A})$, has the following density defined on $\mathbb{R}^N \times \mathbb{R}^N$:

$$p_\epsilon(\boldsymbol{a}, \boldsymbol{x}) = \prod_{i=1}^{N} p_{i,\epsilon}(a^i, x^i), \tag{3.43}$$

where $\boldsymbol{x} = (x^1, \ldots, x^N)$.

Unfortunately, there is no method analogous to the detailed balance equation discussed in Sect. 3.6.1 in the case of the log-linear learning on continuous actions to get the closed form expression for the stationary distribution and study long run properties of the Markov chain from that expression. The technique of general resistance trees [WY15, Fei06, New15], that is useful for the analysis of the independent log-linear learning over discrete actions [MS12], is quite complicated to be applied to the process of independent log-linear learning in continuous action games. That is why in the following a new approach to study Markov chains with continuous states is presented. It turns out that this approach allows us to claim the stochastic stability of the potential function maximizers of the learning procedure under consideration. More precisely, the next theorem can be formulated.

Theorem 3.6.6 *Consider any* continuous *action potential game* $\Gamma = (N, \{A_i\}, \{U_i\}, \phi)$ *with compact action sets* $\{A_i\}$*, continuous utility functions* $\{U_i\}$ *taking values in* $[-1, 0]$*, and the set of potential function maximizers* $\boldsymbol{A}^* = \{\arg\max_{\boldsymbol{a} \in \boldsymbol{A}} \phi(\boldsymbol{a})\}$*. The Markov chain on the space* $\boldsymbol{A} = A_1 \times \cdots \times A_N$ *with the transition probability kernel* $G_\epsilon(\boldsymbol{a}, \cdot)$*,* $\boldsymbol{a} \in \boldsymbol{A}$*, whose density is defined in* (3.43)*, is ergodic for any fixed parameter* ϵ*. Moreover, let* $\Pi_\epsilon(\cdot)$ *be its stationary distribution. Then, given* $m \geq N^2$*, for any* $\varepsilon \in (0, 1]$ *and* $A \in \sigma(\boldsymbol{A})$ *such that* $\boldsymbol{A}^* \cap cl(A) = \emptyset$ *there exists* ϵ' *such that* $\Pi_\epsilon(A) < \varepsilon$ *for all* $\epsilon \leq \epsilon'$*.*

Theorem 3.6.6 claims that by adjusting the algorithm parameter ϵ, one can guarantee that with some given in advance probability $p \in [0, 1)$ the algorithm chooses a fixed neighborhood of the system's optimal states, which under an appropriate game design corresponds to the potential function maximizers in the game.

3.6.2.1 Stochastic Stability of Potential Function Maximizers

The further discussion in this subsection is devoted to the proof of the main result on the continuous version of the independent log-linear learning formulated in Theorem 3.6.6. The idea is to find a sequence $\{G_{\epsilon,n}\}$ of Markov chains that are an approximation for the Markov chain G_ϵ of the learning algorithm introduced in the previous section by (3.41). This sequence should represent an appropriate approximation of the original process. It will allow us to use the results of Appendix A.3.2, in particular Theorem A.3.5, and studying the weak limit of the stationary measures of the chains $\{G_{\epsilon,n}\}$ as $n \to \infty$.

1. **Markov Chain Approximation.** Before constructing an appropriate sequence $\{G_{\epsilon,n}\}$ we prove that the process G_ϵ defined by the independent log-linear learning (3.41) has a unique stationary measure $\Pi_\epsilon(\cdot)$ and $G_\epsilon^t(\boldsymbol{a}, \cdot)$ converges to this measure in total variation for all $\boldsymbol{a} \in \boldsymbol{A}$.

Lemma 3.6.2 *The Markov chain G_ϵ is ergodic with the unique stationary measure Π_ϵ. Moreover, $\lim_{t\to\infty} \|G_\epsilon^t(\boldsymbol{a}, \cdot) - \Pi_\epsilon(\cdot)\|_{\mathrm{TV}} = 0$ for all $\boldsymbol{a} \in \boldsymbol{A}$.*

Proof Firstly, we show that the kernel G_ϵ is $\mathfrak{L}$-irreducible, where $\mathfrak{L}$ is the standard Lebesgue measure on $\boldsymbol{A}$, aperiodic, and recurrent. Let $B \in \sigma(\boldsymbol{A})$, $\mathfrak{L}(B) > 0$. Then, according to (3.42) and (3.43),

$$G_\epsilon(\boldsymbol{a}, B) \geq \epsilon^N \int_B \prod_{i=1}^{N} \frac{\epsilon^{-U_i(\mathbf{z}_i, \boldsymbol{a}^{-i})}}{\int_{A_i} \epsilon^{-U_i(y, \boldsymbol{a}^{-i})} dy} d\mathbf{z}_i > 0.$$

Thus, $G_\epsilon(\boldsymbol{a}, \cdot)$ is $\mathfrak{L}$-irreducible. The recurrence in sense of Definition A.3.2 follows from (3.42), (3.43), and Theorem A.3.2, where the measure $\theta(\cdot)$ is represented by $\mathfrak{L}(\cdot)/\mathfrak{L}(\boldsymbol{A})$. Hence, there exists some unique stationary measure $\Pi_\epsilon(\cdot)$. It implies existence of the unique stationary measure Π_ϵ (see Appendix A.3). Now we show that G_ϵ is aperiodic. Let us suppose that B_1 and B_2 are disjoint non-empty elements of $\sigma(\boldsymbol{A})$ both of positive measure Π_ϵ, with $G_\epsilon(\boldsymbol{a}, B_2) = 1$ for all $\boldsymbol{a} \in B_1$. But it would imply, see again (3.42) and (3.43), that $B_2 = \boldsymbol{A}$, since $G_\epsilon(\boldsymbol{a}, B) < 1$ for any $B \in \sigma(\boldsymbol{A})$: $B \neq \boldsymbol{A}$. Thus, to demonstrate convergence of $G_\epsilon^t(\boldsymbol{a}, \cdot)$ in total variation to Π_ϵ, it remains to prove that the chain is Harris recurrent (see Theorem A.3.4). Indeed, if Π_ϵ is the unique stationary distribution, then $\Pi_\epsilon(A) = \int_{\boldsymbol{A}} G_\epsilon(\boldsymbol{y}, A)\Pi_\epsilon(d\boldsymbol{y})$ for any $A \in \sigma(\boldsymbol{A})$. According to (3.42), $\Pi_\epsilon(A) = 0$, if and only if $\mathfrak{L}(A) = 0$. It is clear that

$\Pr\{\boldsymbol{a}(t) \in A \text{ for all } t|\boldsymbol{a}(0) = \boldsymbol{a}_0\} = 0$ for any $\boldsymbol{a}_0 \in \boldsymbol{A}$ and any $A \in \sigma(\boldsymbol{A}) : \mathfrak{L}(A) = 0$ and $\boldsymbol{a}_0 \notin A$. Moreover, if $A \in \sigma(\boldsymbol{A})$ is such that $\mathfrak{L}(A) = 0$ and $\boldsymbol{a}_0 \in A$, then

$$\Pr\{\boldsymbol{a}(t) \in A \text{ for all } t|\boldsymbol{a}(0) = \boldsymbol{a}_0\} = \lim_{t\to\infty} (1-\epsilon)^t = 0.$$

Thus, according to Definition A.3.5, Harris recurrence follows. □

Now we construct $\{G_{\epsilon,n}\}$, a sequence of transition probability kernels, whose stationary distributions weakly converge to the stationary distribution $\Pi_\epsilon(\cdot)$ of the original Markov chain G_ϵ as $n \to \infty$.

Let us consider a tagged partition $S_n^i = \{(b_n(k_i), c_n(k_i))\}$ of each interval A_i, $i \in [N]$ into n segments with tagged points $a_n(k_i)$, $k_i = 1, \dots, n$. Let each segment of S_n with the tag $\boldsymbol{a}_n(k) = (a_n(k_1), \dots, a_n(k_N))$, k_j is some integer from $\{1, \dots, n\}$, be denoted by $\mathfrak{A}_n(k)$, $k = 1, \dots, n^N$ (see Fig. 3.7).

Now we define two auxiliary Markov chains $G_{\epsilon,n}^{\text{disc}}$ and $G_{\epsilon,n}(\boldsymbol{a}, \cdot)$. The discrete state chain $G_{\epsilon,n}^{\text{disc}}$ is defined on the set of the tags $\boldsymbol{A}_n = \{\boldsymbol{a}_n(k)\}_k$ by the transition probability matrix with the following elements $\{G_{\epsilon,n}^{\text{disc}}(\boldsymbol{a}_n(k), \boldsymbol{a}_n(l))\}_{k,l}$:

$$G_{\epsilon,n}^{\text{disc}}(\boldsymbol{a}_n(k), \boldsymbol{a}_n(l)) = G_\epsilon(\boldsymbol{a}_n(k), \mathfrak{A}_n(l)) = \int_{\mathfrak{A}_n(l)} p_\epsilon(\boldsymbol{a}_n(k), \boldsymbol{x}) d\boldsymbol{x}, \tag{3.44}$$

where p_ϵ is the density function from (3.43). Thus, the elements in the transition probability matrix $G_{\epsilon,n}^{\text{disc}}$ are defined by the probability for the original process G_ϵ to transit from a tag to a segment of the partition S_n. Obviously, the discrete state Markov chain $G_{\epsilon,n}^{\text{disc}}$ is irreducible and aperiodic and, hence, has a unique stationary distribution vector $\boldsymbol{\pi}_{\epsilon,n}^{\text{disc}}$ such that $(G_{\epsilon,n}^{\text{disc}})^t \to \Pi_{\epsilon,n}^{\text{disc}}$, as $t \to \infty$, where $\Pi_{\epsilon,n}^{\text{disc}}$ is the rank one matrix with the rows equal to $\boldsymbol{\pi}_{\epsilon,n}^{\text{disc}}$ (see Theorem A.2.1 in Appendix A.2).

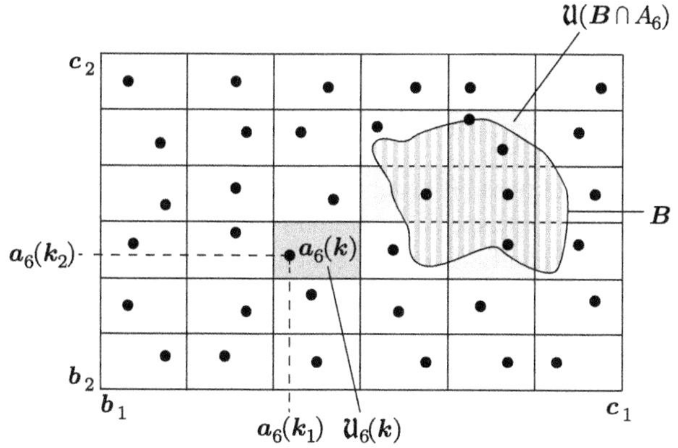

Fig. 3.7 Segmentation of the joint action set, $N = 2$, $n = 6$

The process $G_{\epsilon,n}(\boldsymbol{a},\cdot)$, in its turn, is defined on the whole continuous joint action set $\boldsymbol{A}$ by the following transition probability kernel:

$$G_{\epsilon,n}(\boldsymbol{a},B) = G_\epsilon(\boldsymbol{a},\mathfrak{A}(B\cap \boldsymbol{A}_n)) = \int_{\mathfrak{A}(B\cap \boldsymbol{A}_n)} p_\epsilon(\boldsymbol{a},\boldsymbol{x})d\boldsymbol{x}, \tag{3.45}$$

where p_ϵ is again the density function from (3.43) and $\mathfrak{A}(B\cap \boldsymbol{A}_n) = \bigcup_{l:\boldsymbol{a}_n(l)\in B\cap \boldsymbol{A}_n}\mathfrak{A}_n(l)$. Thus, the definition (3.45) can be reformulated as

$$G_{\epsilon,n}(\boldsymbol{a},B) = \sum_{l:\boldsymbol{a}_n(l)\in B\cap \boldsymbol{A}_n}\int_{\mathfrak{A}_n(l)} p_\epsilon(\boldsymbol{a},\boldsymbol{x})d\boldsymbol{x},$$

which allows us to define the density of $G_{\epsilon,n}$ by the following function:

$$p_{\epsilon,n}(\boldsymbol{a},\boldsymbol{x}) = \sum_{l=1}^{n^N}\delta_{\boldsymbol{a}_n(l)}(\boldsymbol{x})\int_{\mathfrak{A}_n(l)} p_\epsilon(\boldsymbol{a},\mathbf{z})d\mathbf{z}. \tag{3.46}$$

Next, we show that the sequence of the kernels $\{G_{\epsilon,n}\}$ converges to G_ϵ as $n\to\infty$ in sense of Definition A.3.6 (see Appendix A.3.2).

Lemma 3.6.3 *Let S_n be any tagged partition of $\boldsymbol{A}$. Then $G_{\epsilon,n}\to G_\epsilon$ as $n\to\infty$.*

Proof To prove this statement we use Lemma A.3.1. First of all, we show that condition (1) of that lemma is fulfilled for the original process G_ϵ. Let $f:\boldsymbol{A}\to\mathbb{R}$ be a continuous function on $\boldsymbol{A}$. Since $\boldsymbol{A}$ is a compact space, f is bounded on $\boldsymbol{A}$. Then, according to (3.42) and (3.43),

$$\begin{aligned} g(\boldsymbol{x}) &= \int_{\boldsymbol{A}} f(\boldsymbol{y})G_\epsilon(\boldsymbol{x},d\boldsymbol{y}) = \int_{\boldsymbol{A}} f(\boldsymbol{y})p_\epsilon(\boldsymbol{x},\boldsymbol{y})d\boldsymbol{y} \\ &= \int_{\boldsymbol{A}} f(\boldsymbol{y})\sum_{k=0}^{N}\sum_{\nu\in\mathfrak{C}_N^k}(1-\epsilon^m)^k\delta_{x^{\nu(1)}}(y^{\nu(1)})\ldots\delta_{x^{\nu(k)}}(y^{\nu(k)})\epsilon^{m(N-k)} \\ &\quad\times\prod_{j\in[N]\setminus\{\nu(1),\ldots,\nu(k)\}}\frac{\epsilon^{-U_j(y^j,\boldsymbol{x}^{-j})}}{\int_{A_j}\epsilon^{-U_j(z,\boldsymbol{x}^{-j})}dz}d\boldsymbol{y}, \end{aligned}$$

where $\mathfrak{C}_N^k$ is the set of all possible combinations of k elements from the set $[N]$ and $\nu = (\nu(1),\ldots,\nu(k))$ is some special combination from $\mathfrak{C}_N^k$. Taking into account the properties of the Dirac delta function, we conclude that

$$\begin{aligned} g(\boldsymbol{x}) = \sum_{k=0}^{N}\sum_{\nu\in\mathfrak{C}_N^k}\int_{\boldsymbol{A}_{-\nu}}&(1-\epsilon^m)^k f(x^{\nu(1)},\ldots,x^{\nu(k)},\boldsymbol{y}^{-\nu}) \\ &\times\epsilon^{m(N-k)}\prod_{j\in[N]\setminus\{\nu(1),\ldots,\nu(k)\}}\frac{\epsilon^{-U_j(y^j,\boldsymbol{x}^{-j})}}{\int_{A_j}\epsilon^{-U_j(z,\boldsymbol{x}^{-j})}dz}d\boldsymbol{y}^{-\nu}, \end{aligned}$$

where $A_{-\nu} = \times_{j\in[N]\setminus\{\nu(1),\ldots,\nu(k)\}}A_j$. Here, the function $f(x^{\nu(1)},\ldots,x^{\nu(k)},y^{-\nu})$ is the function of $N-k$ variables $y^{-\nu}$ and is equal to the function f with values $x^{\nu(1)},\ldots,x^{\nu(k)}$ on the corresponding arguments y^{ν}. Thus, the function $g(x)$ above is continuous on A, due to continuity of $\{U_i\}$.

Next, it is demonstrated that condition 2) in Lemma A.3.1 is also fulfilled. According to the main property of the weak convergence, it suffices to show that

$$\lim_{n\to\infty}\int_A f(x)G_{\epsilon,n}(a,dx) = \int_A f(x)G_\epsilon(a,dx)$$

for any fixed $a \in A$ and continuous $f : A \to \mathbb{R}$. Using the formula (3.46) for the density of $G_{\epsilon,n}$, we can write

$$\lim_{n\to\infty}\int_A f(x)G_{\epsilon,n}(a,dx) = \lim_{n\to\infty}\sum_{l=1}^{n^N} f(a_n(l))\int_{\mathfrak{A}_n(l)} p_\epsilon(a,z)dz = \int_A f(x)G_\epsilon(a,dx),$$

where the last equality is due to the definition of the Lebesgue integral. □

Thus, according to Theorem A.3.5, it can be concluded that, if $\Pi_{\epsilon,n}(\cdot)$ is a sequence of stationary measures of $G_{\epsilon,n}$, then the unique stationary measure $\Pi_\epsilon(\cdot)$ of G_ϵ is the weak limit of $\Pi_{\epsilon,n}(\cdot)$ as $n\to\infty$. To characterize $\Pi_{\epsilon,n}(\cdot)$ we use the following lemma. It shows that any stationary distribution of $G^{\text{disc}}_{\epsilon,n}$ is also a stationary measure of $G_{\epsilon,n}$.

Lemma 3.6.4 *Let the vector $\boldsymbol{\pi}^{disc}_{\epsilon,n}$ be the unique stationary distribution vector of $G^{disc}_{\epsilon,n}$. Then $\boldsymbol{\pi}^{disc}_{\epsilon,n}$ is also a stationary atomic measure of $G_{\epsilon,n}$.*

Proof Let $\pi^{\text{disc}}_{\epsilon,n}(l)$ be a coordinate of the vector $\boldsymbol{\pi}^{\text{disc}}_{\epsilon,n}$ that is equal to the probability of the chain to be at the state $a_n(l)$ in the stationary distribution. Then the atomic measure $\Pi_{\epsilon,n}(\cdot)$ on A that corresponds to $\boldsymbol{\pi}^{\text{disc}}_{\epsilon,n}$ is defined by the following density function:

$$\pi_{\epsilon,n}(x) = \sum_{l=1}^{n^N}\pi^{\text{disc}}_{\epsilon,n}(l)\delta_{a_n(l)}(x). \tag{3.47}$$

Now we show that $\Pi_{\epsilon,n}G_{\epsilon,n}(B) = \Pi_{\epsilon,n}(B)$ for any $B\in\sigma(A)$, which implies the statement of the lemma. According to (3.47) and since $\sum_{l=1}^{n^N} G^{\text{disc}}_{\epsilon,n}(a_n(l),a_n(k))$ $\pi^{\text{disc}}_{\epsilon,n}(l) = \pi^{\text{disc}}_n(k)$ for any k,

$$\Pi_{\epsilon,n} G_{\epsilon,n}(B) = \int_A G_{\epsilon,n}(\boldsymbol{x}, B)\pi_{\epsilon,n}(\boldsymbol{x})d\boldsymbol{x} = \sum_{l=1}^{n^N} \sum_{k:\boldsymbol{a}_n(k)\in B\cap \boldsymbol{A}_n} G_\epsilon(\boldsymbol{a}_n(l), \mathfrak{A}_n(k))\pi^{\text{disc}}_{\epsilon,n}(l)$$

$$= \sum_{k:\boldsymbol{a}_n(k)\in B\cap \boldsymbol{A}_n} \sum_{l=1}^{n^N} G^{\text{disc}}_{\epsilon,n}(\boldsymbol{a}_n(l), \boldsymbol{a}_n(k))\pi^{\text{disc}}_{\epsilon,n}(l)$$

$$= \sum_{k:\boldsymbol{a}_n(k)\in B\cap \boldsymbol{A}_n} \pi^{\text{disc}}_n(k) = \Pi_{\epsilon,n}(B).$$

□

Further we deal only with those tagged partition, where the set of tags $\boldsymbol{A}_n$ contains some non-empty subset of the potential function maximizers of the game under consideration. Namely, for any n there exists $A^* \subseteq \boldsymbol{A}^*$ such that

$$A^* \subseteq \boldsymbol{A}_n. \tag{3.48}$$

We study the convergence properties of the stationary distribution $\boldsymbol{\pi}^{\text{disc}}_{\epsilon,n}$ of the Markov chain $G^{\text{disc}}_{\epsilon,n}$ defined on such $\boldsymbol{A}_n$ in order to get the characterization for the stationary measure of the process G_ϵ.

2. **Stationary Distribution Vector $\boldsymbol{\pi}^{\text{disc}}_{\epsilon,n}$.** Recall that $G^{\text{disc}}_{\epsilon,n}$ is the Markov chain defined on the finite states $\boldsymbol{A}_n$ represented by the tags of the partition S_n. Let us introduce one more Markov chain $G^{\text{c}}_{\epsilon,n}$ on the same states $\boldsymbol{A}_n$ by $G^{\text{c}}_{\epsilon,n} = P^\epsilon(\boldsymbol{A}_n)$ (see Sect. 3.4.2). Thus, the Markov chain $G^{\text{c}}_{\epsilon,n}$ is defined on the set of tags $\boldsymbol{A}_n$ by the transition probabilities of the discrete independent log-linear learning defined by (3.11), (3.12). According to Theorem 3.4.1, the chain $G^{\text{c}}_{\epsilon,n}$ has the unique stationary distribution. We denote this distribution by $\boldsymbol{\pi}^{\text{c}}_{\epsilon,n}$. Let $\boldsymbol{a}_n(k) = (a_n(k_1), \ldots, a_n(k_N))$ and $\boldsymbol{a}_n(l) = (a_n(l_1), \ldots, a_n(l_N))$ be two tags of the partition S_n. Then by I^{kl}_n we denote the set $\{i : k_i \neq l_i\}$. Now we can use (3.11) to get the following expression:

$$G^{\text{c}}_{\epsilon,n}(\boldsymbol{a}_n(k), \boldsymbol{a}_n(l)) = \sum_{\mathfrak{L}\subseteq\mathfrak{N}: I^{kl}_n\subseteq\mathfrak{L}} \epsilon^{m|\mathfrak{L}|}(1-\epsilon^m)^{|\mathfrak{N}\setminus\mathfrak{L}|} \prod_{j\in\mathfrak{L}} \frac{\epsilon^{-U_j(a_n(l_j), \boldsymbol{a}^{-j}_n(k))}}{\sum_{s_j=1}^{n} \epsilon^{-U_j(a_n(s_j), \boldsymbol{a}^{-j}_n(k))}}. \tag{3.49}$$

According to (3.44),

$$G^{\text{disc}}_{\epsilon,n}(\boldsymbol{a}_n(k), \boldsymbol{a}_n(l)) = \sum_{\mathfrak{L}\subseteq\mathfrak{N}: I^{kl}_n\subseteq\mathfrak{L}} \epsilon^{m|\mathfrak{L}|}(1-\epsilon^m)^{|\mathfrak{N}\setminus\mathfrak{L}|} \prod_{j\in\mathfrak{L}} \frac{\int_{A_n(l_j)} \epsilon^{-U_j(x, \boldsymbol{a}^{-j}_n(k))}dx}{\int_{A_j} \epsilon^{U_j(z, \boldsymbol{a}^{-j}_n(k))}dz}, \tag{3.50}$$

where $\boldsymbol{a}_n^{-j}(k) = (a_n(k_1), \ldots, a_n(k_{j-1}), a_n(k_{j+1}), \ldots, a_n(k_N))$ and $A_n(l_j)$ is the corresponding jth coordinate part of the segment $\mathfrak{A}_n(l)$.

The next lemma claims that the stationary distribution $\boldsymbol{\pi}_{\epsilon,n}^{\text{disc}}$ of the Markov chain $G_{\epsilon,n}^{\text{disc}}$ can be approximated by the stationary distribution $\boldsymbol{\pi}_{\epsilon,n}^{\text{c}}$ of the Markov chain $G_{\epsilon,n}^{\text{c}}$ when n is large.

Lemma 3.6.5 *For any $\varepsilon > 0$, there exists such n' that for any $n \geq n'$ the stationary distributions $\boldsymbol{\pi}_{\epsilon,n}^{\text{disc}}$ and $\boldsymbol{\pi}_{\epsilon,n}^{c}$ of the chains $G_{\epsilon,n}^{\text{disc}}$ and $G_{\epsilon,n}^{c}$ correspondingly satisfy*

$$\|\boldsymbol{\pi}_{\epsilon,n}^{\text{disc}} - \boldsymbol{\pi}_{\epsilon,n}^{c}\| \leq \varepsilon,$$

given $\epsilon < \frac{1}{n}$ and $m \geq N^2$.

For the proof the following result presented in [Jer06] is used.

Theorem 3.6.7 *Let P and $\hat{P}$ be two Markov chains defined on the same finite state space S, $P(s_1, s_2) = p_{s_1 s_2}$, $\hat{P}(s_1, s_2) = \hat{p}_{s_1 s_2}$. Let $\hat{P}^t(s, \cdot)$ be the vector of the s-th row in the transition probability matrix $\hat{P}^t$. Suppose that for some $\delta \geq 0$, $\kappa > 0$*

1. *$\|\hat{P}(s, \cdot) - P(s, \cdot)\|_{l_1} \leq \delta$ for all $s \in S$,*
2. *P is ergodic with a stationary distribution $\boldsymbol{\pi}$ and*

$$\max_{s_1, s_2 \in S} \|P(s_1, \cdot) - P(s_2, \cdot)\|_{l_1} \leq e^{-\kappa},$$

3. *there exists $\varepsilon > 0$ such that $\delta \leq \frac{\varepsilon}{2t_\varepsilon}$, where $t_\varepsilon = \lceil \ln \frac{2}{\varepsilon\kappa} \rceil$.*

Then $\|\hat{P}^t(s, \cdot) - \boldsymbol{\pi}\|_{l_1} \leq \varepsilon$ for any $t \geq t_\varepsilon$ and $s \in S$.

Proof of Lemma 3.6.5 First of all we prove that all conditions of Theorem 3.6.7 are fulfilled for $P = G_{\epsilon,n}^{\text{disc}}$ and $\hat{P} = G_{\epsilon,n}^{\text{c}}$. According to (3.50) and since $\epsilon > 0$,

$$G_{\epsilon,n}^{\text{disc}}(\boldsymbol{a}_n(k), \boldsymbol{a}_n(l)) \leq 1 - \epsilon^m < 1$$

for any tags $\boldsymbol{a}_n(k), \boldsymbol{a}_n(l)$. Hence, there exists $\kappa > 0$ such that for any n

$$\max_{k,l} \|G_{\epsilon,n}^{\text{disc}}(\boldsymbol{a}_n(k), \cdot) - G_{\epsilon,n}^{\text{disc}}(\boldsymbol{a}_n(l), \cdot)\|_{l_1} \leq e^{-\kappa}. \tag{3.51}$$

Further we notice that for any $\varepsilon > 0$ and for any tag $\boldsymbol{a}_n(k)$

$$\|G_{\epsilon,n}^{\text{disc}}(\boldsymbol{a}_n(k), \cdot) - G_{\epsilon,n}^{\text{c}}(\boldsymbol{a}_n(k), \cdot)\|_{l_1} \leq \delta, \tag{3.52}$$

where $\delta \leq \frac{\varepsilon}{2t_\varepsilon}$, $t_\varepsilon = \lceil \ln \frac{2}{\varepsilon\kappa} \rceil$, given sufficiently large n and $\epsilon < \frac{1}{n}$. Indeed, (3.49) and (3.50) imply that for any $\boldsymbol{a}_n(k)$, $k = 1, \ldots, n^N$,

$$\begin{aligned}\|G_{\epsilon,n}^{\text{disc}}(\boldsymbol{a}_n(k), \cdot) - G_{\epsilon,n}^{\text{c}}(\boldsymbol{a}_n(k), \cdot)\|_{l_1} &= \sum_{l=1}^{n^N} |G_{\epsilon,n}^{\text{disc}}(\boldsymbol{a}_n(k), \boldsymbol{a}_n(l)) - G_{\epsilon,n}^{\text{c}}(\boldsymbol{a}_n(k), \boldsymbol{a}_n(l))| \\ &\leq (n^N - 1)\epsilon^m \Delta_n^k + (1 - \epsilon^m)^N \Delta_n^k,\end{aligned} \tag{3.53}$$

where

$$\Delta_n^k = \max_{\mathfrak{L}\subseteq\mathfrak{N}} \prod_{j\in\mathfrak{L}} \left| \frac{\int_{A_n(l_j)} \epsilon^{-U_j(x,a_n^{-j}(k))}dx}{\int_{A_j} \epsilon^{U_j(z,a_n^{-j}(k))}dz} - \frac{\epsilon^{-U_j(a^n(l(j)),a_n^{-j}(k))}}{\sum_{s=1}^n \epsilon^{-U_j(a_n(s(j)),a_n^{-j}(k))}} \right|.$$

Due to the definition of the Riemann integral, Δ_n^k can be done arbitrary small by the choice of a sufficiently large n. Since $m \geq N^2$ and (3.53), $\|G_{\epsilon,n}^{\text{disc}}(\boldsymbol{a}_n(k), \cdot) - G_{\epsilon,n}^{\text{c}}(\boldsymbol{a}_n(k), \cdot)\|_{l_1}$ can be done arbitrary small by the choice of a sufficiently large n and given $\epsilon < \frac{1}{n}$.

Bringing (3.51) and (3.52) together and using Theorem 3.6.7, we conclude that for any $\varepsilon > 0$ there exists n' such that for any $n \geq n'$, $t \geq t_\varepsilon = \lceil \ln \frac{2}{\varepsilon\kappa} \rceil$, and $\epsilon < \frac{1}{n}$

$$\|(G_{\epsilon,n}^{\text{c}})^t(\boldsymbol{a}_n(k), \cdot) - \boldsymbol{\pi}_{\epsilon,n}^{\text{disc}}\|_{l_1} \leq \varepsilon/2.$$

Since $\lim_{t\to\infty}(G_{\epsilon,n}^{\text{c}})^t(\boldsymbol{a}_n(k), \cdot) = \boldsymbol{\pi}_{\epsilon,n}^{\text{c}}$ for any $\boldsymbol{a}_n(k)$, there exists $T \geq t_\varepsilon$ such that

$$\|(G_{\epsilon,n}^{\text{c}})^t(\boldsymbol{a}_n(k), \cdot) - \boldsymbol{\pi}_{\epsilon,n}^{\text{c}}\|_{l_1} \leq \varepsilon/2.$$

for $t \geq T$. Hence,

$$\|\boldsymbol{\pi}_{\epsilon,n}^{\text{disc}} - \boldsymbol{\pi}_{\epsilon,n}^{\text{c}}\|_{l_1} \leq \varepsilon,$$

if the algorithm parameter $\epsilon < \frac{1}{n}$. Thus, the proof is completed. □

Now the following lemma is formulated.

Lemma 3.6.6 *As* $\epsilon \to 0$, $\boldsymbol{\pi}_{\epsilon,n}^{c} \Rightarrow \Pi_{(n)}^*$, *for any* $n \geq 2$ *and* $m \geq N^2$, *where* $\Pi_{(n)}^*$ *is a distribution whose full support is on some subset of* $\boldsymbol{A}^*$.

Proof According to the choice of the tagged partition S_n defined by (3.48), the definition of $G_{\epsilon,n}^{\text{c}}$, and Theorem 3.4.1, we conclude that for any $n \geq 2$ the stochastic stable states in $\boldsymbol{\pi}_{\epsilon,n}^{\text{c}}$ are contained in the set of potential function maximizers. Thus, the result follows. □

3. Proof of Theorem 3.6.6.

Proof According to Lemma 3.6.3, Theorem A.3.5, and Lemma 3.6.4, the stationary distribution Π_ϵ of the Markov chain G_ϵ is the weak limit of the stationary distributions $\boldsymbol{\pi}_{\epsilon,n}^{\text{disc}}$ of the Markov chain $G_{\epsilon,n}^{\text{disc}}$. It implies that for any $\varepsilon > 0$ there exists $n_0(\varepsilon)$ such that

$$\left\|\boldsymbol{\pi}_{\epsilon,n}^{\text{disc}} - \Pi_\epsilon\right\|_w \leq \varepsilon/4 \tag{3.54}$$

for any $n \geq n_0$. Since $m \geq N^2$ and the total variation convergence is equivalent to the weak convergence for discrete measures (see Appendix A.1), Lemma 3.6.5 can be used to conclude that there exists $n' = n'(\varepsilon)$ such that

$$\left|\int_A f(\boldsymbol{x})\boldsymbol{\pi}^{\text{disc}}_{\epsilon,n}(dx) - \int_A f(\boldsymbol{x})\boldsymbol{\pi}^{\text{c}}_{\epsilon,n}(dx)\right| \leq \varepsilon/4$$

for any $n \geq n'$ and $\epsilon < \frac{1}{n}$. Hence, (3.54) and the inequality above imply

$$\left|\int_A f(\boldsymbol{x})\Pi_\epsilon(dx) - \int_A f(\boldsymbol{x})\boldsymbol{\pi}^{\text{c}}_{\epsilon,n}(dx)\right| \leq \varepsilon/2 \tag{3.55}$$

for any $n \geq n_1(\varepsilon) = \max(n_0, n')$ and $\epsilon < \frac{1}{n}$. According to Lemma 3.6.6, there exists $\epsilon'(\varepsilon)$ such that for any $\epsilon \leq \epsilon'$ and $n \geq 2$

$$\left|\int_A f(\boldsymbol{x})\boldsymbol{\pi}^{\text{c}}_{\epsilon,n}(dx) - \int_A f(\boldsymbol{x})\Pi^*_{(n)}(dx)\right| \leq \varepsilon/2. \tag{3.56}$$

Thus, taking into account (3.55) and (3.56), we conclude that for any $\varepsilon > 0$ there exists $\tilde{\epsilon}(\varepsilon)$, $\tilde{\epsilon} = \min\left(\frac{1}{n_1(\varepsilon)}, \epsilon'\right)$, such that for any $n \geq n_1(\varepsilon)$

$$\left|\int_A f(\boldsymbol{x})\Pi_\epsilon(dx) - \int_A f(\boldsymbol{x})\Pi^*_{(n)}(dx)\right| \leq \varepsilon$$

for any $\epsilon \leq \tilde{\epsilon}$, which implies the claim of the theorem. □

3.6.3 *Choice of the Parameter*

The technique of Markov chain approximation has been used to show that for any given in advance probability $p < 1$ there exists such parameter $\epsilon > 0$ that the independent log-linear learning with the parameter ϵ chooses a fixed neighborhood of the optimal Nash equilibria with the probability p as time tends to infinity. To get a higher probability p in long run, one has to set up ϵ small enough. Thus, if ϵ is small, and according to the definition of the learning algorithm (3.41), convergence of the corresponding chain to its unique stationary distribution can be very slow. Recall that each agent explores by changing her action with the probability ϵ^m at each step ($m \geq N^2$).

To rectify this issue, this subsection presents analysis of finite time behavior of the time-inhomogeneous Markov chain obtained by settings, where the parameter ϵ in the independent log-linear learning is a time-dependent function tending to zero as time runs. Although convergence of this chain is not claimed in any sense, the following proposition asserts that a polynomial parameter $\epsilon(t)$ can be chosen to guarantee convergence to an optimal Nash equilibrium with some given in advance probability after some finite number of steps. Analogously to the case of discrete action games (see Theorem 3.5.4), an exponential decrease of the parameter $\epsilon(t)$ to 0 is set up. Note that under this choice of the exploration parameter ϵ, and given $\beta = -\ln\epsilon$, the rationality parameter β of the independent log-linear algorithm is a polynomial function increasing to infinity as time runs.

Theorem 3.6.8 *Let $\Gamma = (N, \{A_i\}_i, \{U_i\}_i, \phi)$ be a* continuous *action potential game with compact action sets $\{A_i\}$, continuous utility functions $\{U_i\}$ taking values in $[-1, 0]$, and the set of potential function maximizers $\boldsymbol{A}^* = \{\arg\max_{\boldsymbol{a}\in\boldsymbol{A}} \phi(\boldsymbol{a})\}$. Suppose that each player uses the independent log-linear learning* (3.41) *with the time-dependent parameter $\epsilon(S,t) = \frac{S}{\exp\{(t+1)^\alpha\}}$, where α is fixed. Then for any $\varepsilon \in (0,1)$ there exist such constants T and $S(T)$ that $\|G^t_{\epsilon(S(T),t)}(\boldsymbol{a},\cdot) - \Pi^*(\cdot)\|_w \le \varepsilon$ for any $t \ge T$ and any $\boldsymbol{a} \in \boldsymbol{A}$, where Π^* is some distribution with full support on some subset of $\boldsymbol{A}^*$.*

Proof Note that, according to Theorem 3.6.6, there exists such T_1 that

$$\|\Pi_{\epsilon(S,t)} - \Pi^*_{(t)}\|_w < \varepsilon/3 \tag{3.57}$$

for any $S > 0$, $t \ge T_1$ and given $\varepsilon > 0$, where $\Pi^*_{(t)}$ is a distribution with full support on a subset of the set $\boldsymbol{A}^*$ for any t. Using the triangle inequality, we can write:

$$\begin{aligned}\|G^t_{\epsilon(S,t)} - \Pi^*_{(t)}\|_w &\le \|G^{k-1}_{\epsilon(S,t)} G^{k,t}_{\epsilon(S,t)} - \Pi_{\epsilon(S,k)} G^{k,t}_{\epsilon(S,t)}\|_w \\ &+ \|\Pi_{\epsilon(S,k)} G^{k,t}_{\epsilon(S,t)} - \Pi_{\epsilon(S,t)}\|_w + \|\Pi_{\epsilon(S,t)} - \Pi^*_{(t)}\|_w,\end{aligned} \tag{3.58}$$

for any t and k: $k < t$.

Now we turn to the second term in (3.58). Using the reasoning analogous to one in (3.23) of Theorem 3.5.3, we conclude that

$$\|\Pi_{\epsilon(S,k)} G^{k,t}_{\epsilon(S,t)} - \Pi_{\epsilon(S,t)}\|_w \le \sum_{j=k}^{t} \|\Pi_{\epsilon(S,j)} - \Pi_{\epsilon(S,j+1)}\|_w. \tag{3.59}$$

Let us consider any $t > T_1$. Taking into account (3.54), we conclude that for any ϵ and given $\varepsilon > 0$ there exists $n = n(t,k,\varepsilon)$ such that $\|\Pi_\epsilon - \boldsymbol{\pi}^{\mathrm{disc}}_{\epsilon,n}\|_w \le \frac{\varepsilon}{12(t-k+1)}$. Using the triangle inequality again, we get

$$\begin{aligned}\|\Pi_{\epsilon(S,j)} - \Pi_{\epsilon(S,j+1)}\|_w &\le \|\Pi_{\epsilon(S,j)} - \boldsymbol{\pi}^{\mathrm{disc}}_{\epsilon(S,j),n}\|_w + \|\Pi_{\epsilon(S,j+1)} - \boldsymbol{\pi}^{\mathrm{disc}}_{\epsilon(S,j+1),n}\|_w \\ &+ \|\boldsymbol{\pi}^{\mathrm{disc}}_{\epsilon(S,j),n} - \boldsymbol{\pi}^{\mathrm{disc}}_{\epsilon(S,j+1),n}\|_{l_1} \\ &\le \frac{\varepsilon}{6(t-k+1)} + \|\boldsymbol{\pi}^{\mathrm{disc}}_{\epsilon(S,j),n} - \boldsymbol{\pi}^{\mathrm{disc}}_{\epsilon(S,j+1),n}\|_{l_1}.\end{aligned} \tag{3.60}$$

Further, we note that, according to (3.44), the chain $G^{\text{disc}}_{\epsilon,n}$ is a regular perturbed processes in respect to the parameter ϵ, namely

$$G^{\text{disc}}_{\epsilon,n}(\boldsymbol{a}_n(k), \boldsymbol{a}_n(l)) = O(\epsilon^{R(\boldsymbol{a}_n(k),\boldsymbol{a}_n(l))}),$$

where $R(\boldsymbol{a}_n(k), \boldsymbol{a}_n(l))$ is some nonnegative constant called the resistance for the states $\boldsymbol{a}_n(k)$ and $\boldsymbol{a}_n(l)$. Thus, according to Theorem A.2.2 (see Appendix A.2.1), each coordinate in the stationary distribution vector $\boldsymbol{\pi}^{\text{disc}}_{\epsilon,n}$ is of type $O(f(\epsilon))$, where $f(\epsilon)$ is some rational function of ϵ. It allows us to use the argumentation analogous to one in Theorem 3.2.2 to conclude that

$$\sum_{j=1}^{\infty} \|\boldsymbol{\pi}^{\text{disc}}_{\epsilon(S,j),n} - \boldsymbol{\pi}^{\text{disc}}_{\epsilon(S,j+1),n}\|_{l_1} < \infty.$$

Hence, there exists $T_2 \geq T_1$ and $k(T_2)$ such that $\sum_{j=k}^{t} \|\boldsymbol{\pi}^{\text{disc}}_{\epsilon(S,j),n} - \boldsymbol{\pi}^{\text{disc}}_{\epsilon(S,j+1),n}\|_{l_1} < \frac{\varepsilon}{6}$ for any $S > 0$, $t \geq T_2$, and $k(T_2) \leq k \leq T_2$. Thus, taking into account (3.59) and (3.60), we get

$$\|\Pi_{\epsilon(S,k)} G^{k,t}_{\epsilon(S,t)} - \Pi_{\epsilon(S,t)}\|_w \leq \frac{\varepsilon}{3} \tag{3.61}$$

for any finite $t : t \geq T_2$.

To bound the first term of the inequality (3.58), we can use the inequality (see Lemma A.3.2)

$$\|G^{k-1}_{\epsilon(S,t)} G^{k,t}_{\epsilon(S,t)} - \Pi_{\epsilon(S,k)} G^{k,t}_{\epsilon(S,t)}\|_{\text{TV}} \leq \tau(G^{k,t}_{\epsilon(S,t)}),$$

where $\tau(G^{k,t}_{\epsilon(S,t)})$ is the coefficient of ergodicity for the kernel $G^{k,t}_{\epsilon(S,t)}$ (see Definition A.3.7 in Appendix A.3.3). According to this definition of the coefficient of ergodicity and definition of the Markov chain G_ϵ, $\tau(G_{\epsilon(S,t)})$ is strictly less than 1. More precisely, using definition of the algorithm (3.41) and since $U_i \in [-1, 0]$ for all $i = 1, \dots, N$, we conclude that there exist such $\boldsymbol{a}_0 \in \boldsymbol{A}$ and $A_0 \in \sigma(\boldsymbol{A})$ that

$$\tau(G_{\epsilon(S,t)}) \leq 1 - G_{\epsilon(S,t)}(\boldsymbol{a}_0, A_0) \leq 1 - K\epsilon^{mN+N},$$

where K is some positive constant. Thus, according to Lemma A.3.3 in Appendix A.3.3, (A.1) in Appendix A.1, and given $\epsilon(S,t) = \frac{S}{\exp\{(t+1)^\alpha\}}$, we get

$$\|G^{k-1}_{\epsilon(S,t)} G^{k,t}_{\epsilon(S,t)} - \Pi_{\epsilon(S,k)} G^{k,t}_{\epsilon(S,t)}\|_w \leq \tau(G^{k,t}_{\epsilon(S,t)}) \leq \left(1 - K\frac{S^{mN+N}}{\exp\{(t+1)^\alpha(mN+N)\}}\right)^{t-k}.$$

Note that for T_2 and any $\varepsilon \in (0, 1)$ there exists such $S = S(T_2)$ that

$$S^{mN+N} > \frac{1}{K}\left(1 - \left(\frac{\varepsilon}{3}\right)^{\frac{1}{T_2-k}}\right) \exp\{(T_2 + 1)^{\alpha}(mN + N)\}.$$

This implies that

$$\tau(G^{k,T_2}_{\epsilon(S,t)}) \leq \left(1 - K\frac{S}{\exp(t + 1)^{\alpha}(mN + N)}\right)^{T_2-k} < \frac{\varepsilon}{3},$$

and, hence,

$$\|G^{k-1}_{\epsilon(S,t)} G^{k,t}_{\epsilon(S,t)} - \Pi_{\epsilon(S,k)} G^{k,t}_{\epsilon(S,t)}\|_w \leq \tau(G^{k,t}_{\epsilon(S,t)}) \leq \tau(G^{k,T_2}_{\epsilon(S,t)}) < \varepsilon/3 \tag{3.62}$$

for $t \geq T_2$, where the second inequality is due to Lemma A.3.3 in Appendix A.3.3.

Thus, bringing (3.58), (3.57), (3.61), and (3.62) together, we conclude that

$$\|G^{t}_{\epsilon(S,t)}(\boldsymbol{a}, \cdot) - \Pi^{*}_{(t)}(\cdot)\|_w \leq \varepsilon$$

for any given $\varepsilon > 0$ and for any finite $t \geq T = T_2$. Taking into account that $\Pi^{*}_{(t)}(\cdot)$ is a distribution over a subset of the set $\boldsymbol{A}^*$, we conclude the proof. □

Thus, the theorem above "justifies" a time-dependent choice of the parameter in the independent log-linear learning in continuous action potential games.

Remark 3.6.2 It is worth noting that the proof of the theorem above can be also applied to the asynchronous log-linear learning to obtain the analogous result on finite time behavior for that procedure. In other words, a polynomial time-dependent parameter $\beta(t)$ can be set up in such a way that the Markov chain $P_{\beta(t)}$ introduced in Sect. 3.6.1 and defined by the kernel (3.26) achieves an optimal state with a high given in advance probability after some finite time. The simulation results presented below support these choices of the parameters in both cases of the asynchronous and synchronous independent log-linear learning.

3.6.4 Simulation Results: Example of a Routing Problem

In this section, a distributed routing problem introduced in Example 2.2.3 is designed. We consider a network system consisting of five nodes $[M] = \{A, B, C, D, E\}$ and five links $L = \{l_1 = AB, l_2 = BB_1C, l_3 = BB_2C, l_4 = CD, l_5 = CE\}$ (see Fig. 3.8). There are five different routes (paths) over this network, namely $R_1 = l_1 \to l_2 \to l_4$ (blue), $R_2 = l_1 \to l_3 \to l_5$ (red), $R_3 = l_2 \to l_5$ (brown), $R_4 = l_3 \to l_4$ (orange), $R_5 = l_4 \to l_5$ (green). The link costs are defined by the following functions: $P_{l_1}(x) = 2.5x^2 + x$, $P_{l_2}(x) = 4x^2 + 2x + 1$, $P_{l_3}(x) = 5x^2$, $P_{l_4}(x) = 2x^2 + x$, $P_{l_5}(x) = 10x$. We consider N users (agents) who are adhered to

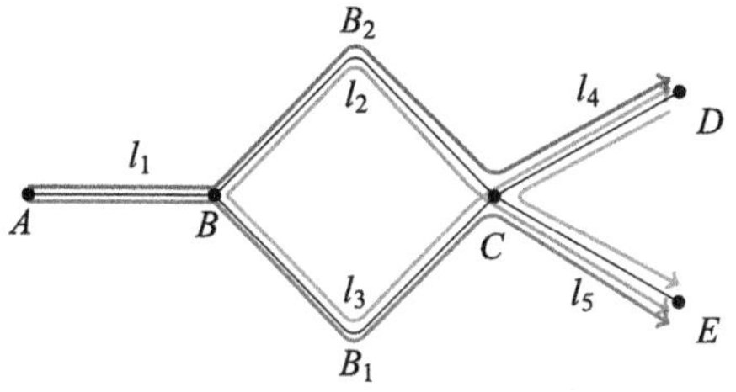

Fig. 3.8 Network with five routes

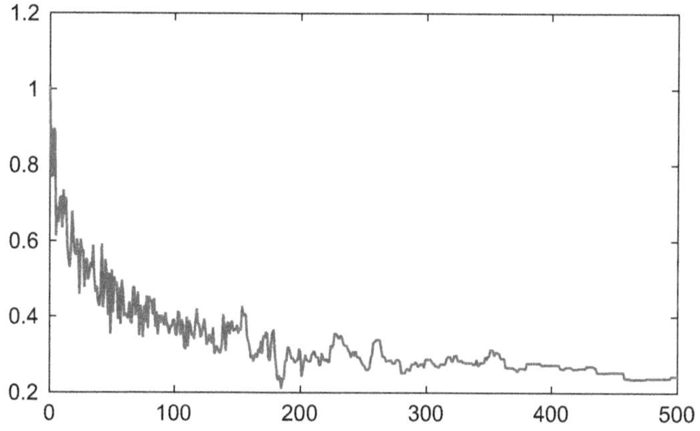

Fig. 3.9 The value $\frac{\|x(t)-x^*\|}{\|x^*\|}$ (y-axis) in the run of the algorithm (x-axis), $N = 20$

one of the paths from the set $\{R_1, R_2, R_3, R_4, R_5\}$. The agents need to decide on the flow $x_i \in X_i = [0, x_{i,\max}]$, $i \in [N]$, to be sent over her route. The local profit functions of the agents are $u_i(x_i) = \alpha_i x_i^3 + \beta_i x_i^2 + \gamma_i x_i + \tau_i$, where $\alpha_i, \beta_i, \gamma_i$, and τ_i are some nonnegative scalars. Thus, the global optimization problem in this network system is to maximize the function $\phi : \mathbb{R}^N \to \mathbb{R}$:

$$\phi(\boldsymbol{x}) = \phi(x_1, \ldots, x_N) = \sum_{i=1}^{N} u_i(x_i) - \sum_{k=1}^{5} P_{l_k}\left(\sum_{j:l_k \in R_j} x_j\right).$$

According to discussion in Example 2.2.3, the game $\Gamma(N, \boldsymbol{X}, \{U_i\}, \phi)$ is potential, given that the local utility functions $\{U_i\}$ are defined as follows:

$$U_i(\boldsymbol{x}) = u_i(x_i) - \sum_{k:l_k \in R_i} P_l\left(\sum_{j:l_k \in R_j} x_j\right),$$

where $R_i \in \{R_1, R_2, R_3, R_4, R_5\}$ is the path in the network corresponding to the route of the agent i.

Figures 3.9, 3.10, and 3.11 show the performance of the independent log-linear learning with $\epsilon(t) = 10\exp(-0.1t^{0.7})$ after 500 iterations for different number of

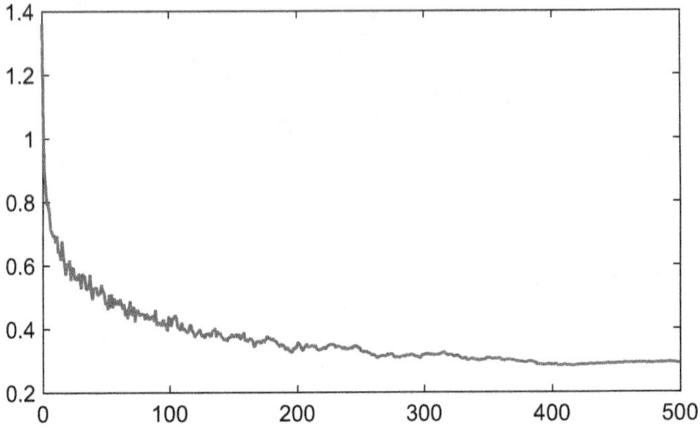

Fig. 3.10 The value $\frac{\|x(t)-x^*\|}{\|x^*\|}$ (y-axis) in the run of the algorithm (x-axis), $N = 200$

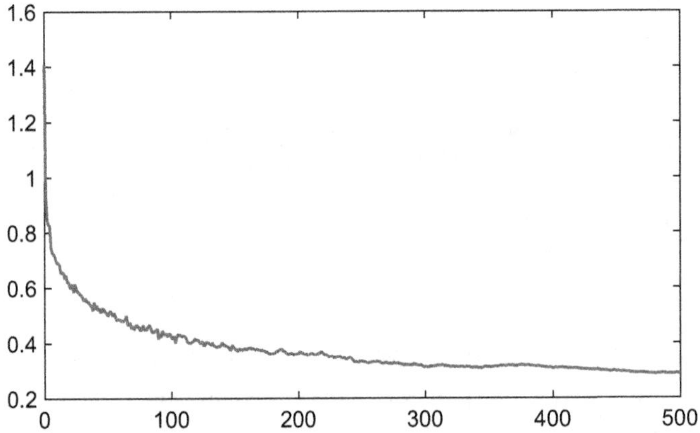

Fig. 3.11 The value $\frac{\|x(t)-x^*\|}{\|x^*\|}$ (y-axis) in the run of the algorithm (x-axis), $N = 2000$

players: $N = 20$, $N = 200$, and $N = 2000$. The points on the figures represent the value $\frac{\|x(t)-x^*\|}{\|x^*\|}$ at each moment of time t. This value indicates the relative error of the chosen joint action $\boldsymbol{x}(t)$ in respect to a Nash equilibrium $\boldsymbol{x}^*$ that maximizes the global objective function $\phi(\boldsymbol{x})$. According to the implementation (see Fig. 3.9), the relative error is 25% after 500 steps in the case $N = 20$. We can see that the relative error is reduced to 30% already after 100 iterations. The further decrease is not rapid, since the low exploration rate already after 200 iterations ($\epsilon(t = 200) = 0.17$, $\epsilon(t = 300) = 0.04$). If the number of the users is increased to $N = 200$ and $N = 2000$, the run of the algorithm changes insignificantly. The corresponding graphs (see Figs. 3.10 and 3.11) demonstrate that the relative error 30% is achieved after 500 iterations. Similarly to the case of 20 users, the joint action continues approaching an optimal Nash equilibrium during last 200 iterations. However, due

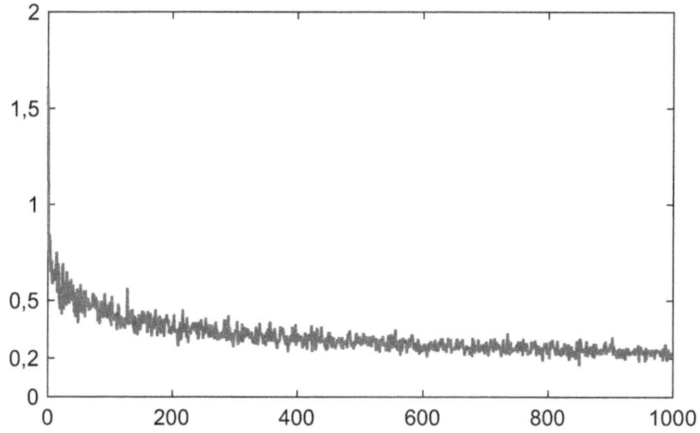

Fig. 3.12 The value $\frac{\|x(t)-x^*\|}{\|x^*\|}$ (*y*-axis) in the run of the algorithm (*x*-axis), $N = 2000$

to the low values of the exploration rate at that steps, the decrease of the relative error is slow. Moreover, by adjusting the exploration rate, namely by setting $\epsilon(t) = \exp(-0.1t^{0.5})$, one can keep the relative error decreasing further on. Figure 3.12 shows that the algorithm with $N = 2000$ players can reduce the relative error to 20% after 1000 iterations.

Although the decrease of the relative error is not rapid, one can still observe that under an appropriate setting for the exploration parameter the algorithm approaches an efficient Nash equilibrium as time runs. Moreover, there is no significant difference in the relative errors, if the number of players has different orders. Thus, we can empirically conclude that the independent log-linear algorithm is scalable in respect to the number of agents in the system. However, the explicit relation between the approach rate of the algorithm, given the setting in Theorem 3.6.8, the game dimension, and the choice of the parameter $\epsilon(t)$ needs to be additionally investigated in the future work.

3.7 Conclusion

This chapter presents the memoryless learning algorithms based on logit dynamics, which can be applied to systems with access to the oracle-based information. Figure 3.13 contains the summary of the results regarding this learning procedures.

Firstly, the chapter deals with discrete action potential games. Here some sufficient conditions for a general efficient memoryless learning, which is defined by a regular perturbed Markov chain, are formulated (Theorem 3.2.2). Further, the log-linear learning and independent log-linear learning are demonstrated to fulfill these conditions and, thus, under an appropriate choice of the algorithm parameters, converge in total variation to the distribution with full support on potential function

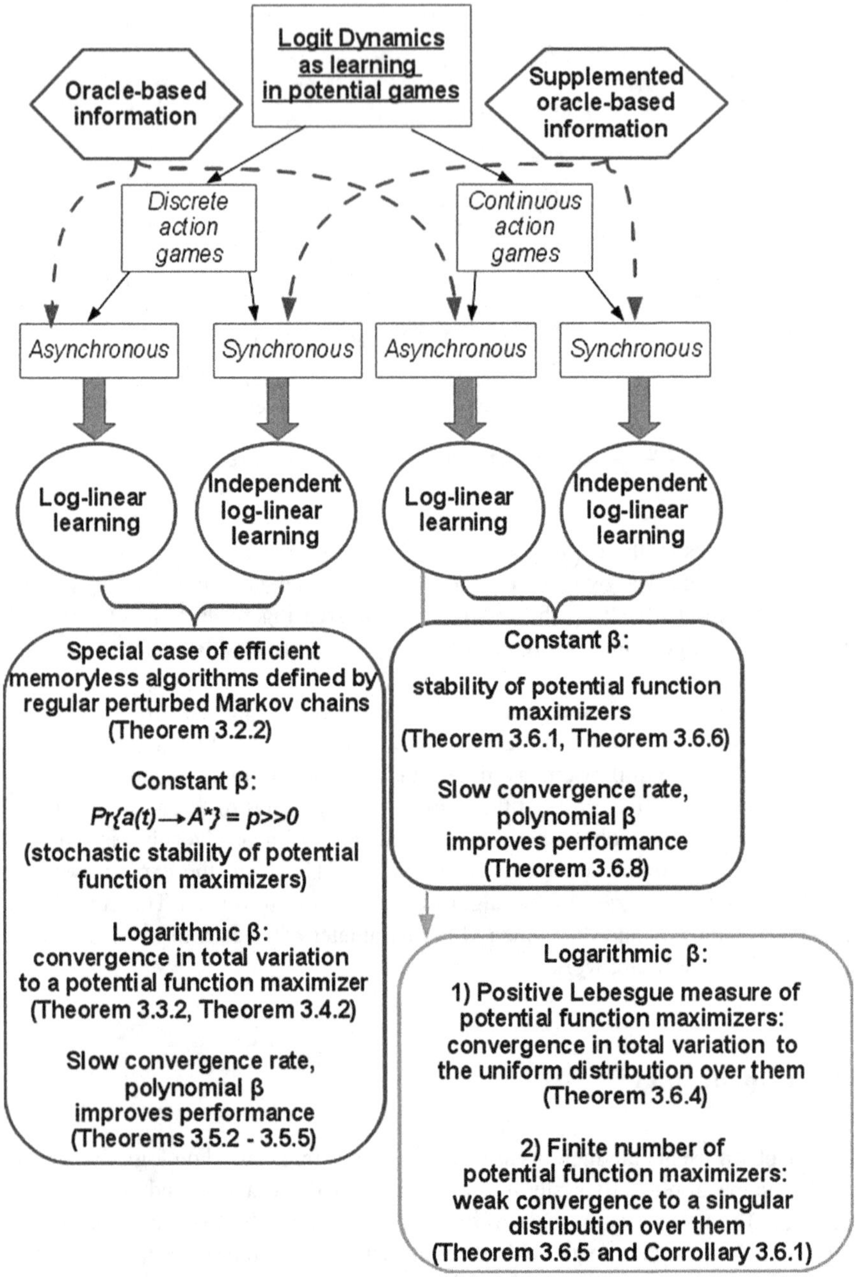

Fig. 3.13 Summary of the results

maximizers coinciding with the system's optimal states. The corresponding settings for the algorithms' parameters are also investigated. The logarithmic rationality parameter guarantees desired convergence (Theorems 3.3.2 and 3.4.2). However, the logarithmic function increases very slowly. It means players need a lot of time to achieve an appropriate level of the rationality and, thus, the algorithms have a low convergence rate under the logarithmic parameter setting (Theorems 3.5.2 and 3.5.4). The analysis of finite time behavior, however, proposes an alternative choice of the polynomial parameter (Theorems 3.5.3 and 3.5.5). The simulation results demonstrate the effectiveness of such setting. The main difference between the log-linear learning and independent log-linear learning is their asynchronous and synchronous performance, respectively. It is also demonstrated that the synchronization is impossible in the general case given the pure oracle-based information available in the system (Sect. 3.4.1). That is why to organize the synchronization efficiently one needs to provide players with some addition information. In the case of the independent log-linear learning this information is the players' knowledge about their currently played actions (supplemented oracle-based information).

The second part of the chapter is devoted to learning in continuous action potential games. The discrete log-linear and independent log-linear learning procedures are extended to this case. In contrast to the case of discrete action games, the Markov chain of the independent log-linear learning with continuous states cannot be studied by the theory of resistance trees. That is why the new approach is developed to analyze long run properties of this chain with a time-invariant setting and to prove definite stability of the potential function maximizers in the correspond learning algorithm (Theorem 3.6.6). For the log-linear learning not only stochastic stability of the optimal states is demonstrated, but also the time-dependent logarithmic parameter is found that guarantees the total variation and weak convergence of the learning procedure to a distribution over optima, if the potential function maximizers have some positive and zero Lebesgue measure, respectively (Theorems 3.6.4 and 3.6.5). Moreover, analogously to the case of discrete actions, the polynomial rationality parameter can be chosen to improve the performance of the algorithms (Theorem 3.6.8).

References

[AFN08] C. Alós-Ferrer, N. Netzer, The logit-response dynamics. TWI Research Paper Series 28, Thurgauer Wirtschaftsinstitut, Universitaet Konstanz (2008)

[AFPP12] V. Auletta, D. Ferraioli, F. Pasquale, G. Persiano, Metastability of logit dynamics for coordination games, in *Proceedings of the Twenty-Third Annual ACM-SIAM Symposium on Discrete Algorithms, SODA '12*, (SIAM, Providence, RI, 2012), pp. 1006–1024

[AMS07] G. Arslan, J.R. Marden, J.S. Shamma, Autonomous vehicle-target assignment: a game theoretical formulation. ASME J. Dyn. Syst. Meas. Control **129**, 584–596 (2007)

[Blu93] L.E. Blume, The statistical mechanics of strategic interaction. Games Econ. Behav. **5**(3), 387–424 (1993)

[BM59] H.D. Block, J. Marschak, Random orderings and stochastic theories of response. Cowles Foundation Discussion Papers 66, Cowles Foundation for Research in Economics, Yale University (1959)

[CT06] T.M. Cover, J.A. Thomas, *Elements of Information Theory*. Wiley Series in Telecommunications and Signal Processing (Wiley-Interscience, New York, 2006)

[Ell93] G. Ellison, Learning, local interaction, and coordination. Econometrica **61**(5), 1047–1071 (1993)

[Fei06] Y. Feinberg, Evolutionary dynamics and long-run selection. B.E. J. Theor. Econ. **6**(1), 1–26 (2006)

[FL95] D. Fudenberg, D.K. Levine, Consistency and cautious fictitious play. J. Econ. Dyn. Control **19**(5–7), 1065–1089 (1995)

[FL99] D. Fudenberg, D.K. Levine, Conditional universal consistency. Scholarly articles, Harvard University Department of Economics (1999)

[HMC06] S. Hart, A. Mas-Colell, Stochastic uncoupled dynamics and Nash equilibrium. Games Econ. Behav. **41**(1), 286–303 (2006)

[HS91] H. Haario, E. Saksman, Simulated annealing process in general state space. Adv. Appl. Prob. **23**, 866–893 (1991)

[Hwa80] C.-R. Hwang, Laplace's method revisited: weak convergence of probability measures. Ann. Probab. **8**(6), 1177–1182 (1980)

[Jer06] M. Jerrum, On the approximation of one Markov chain by another. Probab. Theory Relat. Fields **135**(1), 1–14 (2006)

[LS13] Y. Lim, J.S. Shamma, Robustness of stochastic stability in game theoretic learning, in *2013 American Control Conference*, June 2013, pp. 6145–6150

[MAS09a] J.R. Marden, G. Arslan, J.S. Shamma, Cooperative control and potential games. Trans. Sys. Man Cyber. B **39**(6), 1393–1407 (2009)

[MBML12] P. Mertikopoulos, E.V. Belmega, A. Moustakas, S. Lasaulce, Distributed learning policies for power allocation in multiple access channels. IEEE J. Sel. Areas Commun. **30**(1), 1–11 (2012)

[McF74] D. McFadden, Conditional logit analysis of qualitative choice behavior, in *Frontiers in Econometrics*, ed. by P. Zarembka (Academic, New York, 1974), pp. 105–142

[MGPS07] F. Meshkati, A.J. Goldsmith, H.V. Poor, S.C. Schwartz, A game-theoretic approach to energy-efficient modulation in CDMA networks with delay QoS constraints. IEEE J. Sel. A. Commun. **25**(6), 1069–1078 (2007)

[MRSV85] D. Mitra, F. Romeo, A. Sangiovanni-Vincentelli, Convergence and finite-time behavior of simulated annealing, in *24th IEEE Conference on Decision and Control (CDC), 1985*, December 1985, pp. 761–767

[MS09] A. Montanari, A. Saberi, Convergence to equilibrium in local interaction games. SIGecom Exch. **8**(1), 11:1–11:4 (2009)

[MS12] J.R. Marden, J.S. Shamma, Revisiting log-linear learning: asynchrony, completeness and payoff-based implementation. Games Econ. Behav. **75**(2), 788–808 (2012)

[New15] J. Newton, Stochastic stability on general state spaces. J. Math. Econ. **58**, 46–60 (2015)

[PML15] S. Perkins, P. Mertikopoulos, D.S. Leslie, Mixed-strategy learning with continuous action sets. IEEE Trans. Autom. Control **62**, 379–384 (2015)

[Pro56] Y.V. Prokhorov, Convergence of random processes and limit theorems in probability theory. Theory Probab. Appl. **1**(2), 157–214 (1956)

[Ros73] R.W. Rosenthal, A class of games possessing pure-strategy Nash equilibria. Int. J. Game Theory **2**, 65–67 (1973)

[RR04] G.O. Roberts, J.S. Rosenthal, General state space Markov chains and MCMC algorithms. Probab. Surv. **1**, 20–71 (2004)

[SPB08] G. Scutari, D.P. Palomar, S. Barbarossa, Competitive design of multiuser MIMO systems based on game theory: a unified view. IEEE J. Sel. Areas Commun. **26**(7), 1089–1103 (2008)

[SZPB12] W. Saad, H. Zhu, H.V. Poor, T. Basar, Game-theoretic methods for the smart grid: an overview of microgrid systems, demand-side management, and smart grid communications. IEEE Signal Process. Mag. **29**(5), 86–105 (2012)

[Tat14a] T. Tatarenko, Log-linear learning: convergence in discrete and continuous strategy potential games, in *53rd IEEE Conference on Decision and Control*, December 2014, pp. 426–432

[Tat14b] T. Tatarenko, Proving convergence of log-linear learning in potential games, in *American Control Conference (ACC), 2014*, June 2014, pp. 972–977

[Tat16b] T. Tatarenko, Stochastic stability of potential function maximizers in continuous version of independent log-linear learning, in *European Control Conference (ECC), 2016*, June 2016, pp. 210–215

[WY15] C. Wallace, H.P. Young, Stochastic evolutionary game dynamics, in *Handbook of Game Theory with Economic Applications*, Chap. 6, vol. 4 (Elsevier, Amsterdam, 2015), pp. 327–380

[You02] P.H. Young, The diffusion of innovations in social networks. Technical report (2002)

[You93] H.P. Young, The evolution of conventions. Econometrica **61**(1), 57–84 (1993)

[ZM13b] M. Zhu, S. Martínez, Distributed coverage games for energy-aware mobile sensor networks. SIAM J. Control Optim. **51**(1), 1–27 (2013)

Chapter 4
Stochastic Methods in Distributed Optimization and Game-Theoretic Learning

4.1 Introduction

This chapter studies the way to apply stochastic approximation procedure, known as the Robbins–Monro procedure [RM51], to *distributed non-convex optimization* as well as to *communication-* and *payoff-based learning in potential games*. The motivation of this analysis is as follows. In the seminal work [NK73] the authors study the Robbins–Monro procedure in terms of iterative Markov processes. They demonstrate that the procedure with some specific setting converges almost surely to a local optimum, being applied to a general centralized *non-convex* optimization problem. Since applications of *distributed* optimization have gained a lot of popularity recently [PE09], it is of great importance to investigate the possibility to "decentralize" the algorithm with the same efficient performance.

Firstly, we focus on *push-sum* based distributed optimization. The push-sum algorithm is initially introduced in [KDG03] for computation of aggregate information and used in [TLR12a] for distributed optimization. The work [NO15] studies this algorithm over *time-varying communication* in the case of well-behaved convex functions. The papers [NO15, NO14] demonstrate convergence of this algorithm to the unique optimum in the case of a convex objective function. Moreover, the papers [NO15, NO14] study the convergence rate of the procedure and, thus, provide some important insights into the algorithm's performance. Analogously to other algorithms applicable to distributed optimization, the push-sum algorithm is based on a two-stage procedure. On the first stage agents exchange the information about their current states according to the specified communication protocol. After that the local gradient descent step is adapted to move each agent toward a minimum of her objective function. The main advantage of the push-sum algorithm consists in utilization of such communication protocol that enables us to overcome the restrictive assumptions on the graph structure. More precisely, this algorithm requires no knowledge of either the link weights or the number of agents and removes the need for the double stochastic communication matrices. Thus, this approach can be

T. Tatarenko, *Game-Theoretic Learning and Distributed Optimization in Memoryless Multi-Agent Systems*, DOI 10.1007/978-3-319-65479-9_4

applied to network optimization with *time-varying directed communication*, where the connection between nodes depends on the broadcast range and nodes are able to broadcast messages at different power levels rending unidirectional links [NO15].

So far, most of the theoretical work on the topic of distributed optimization has been devoted to optimization of a sum of well-behaved *convex functions*, where the assumption on subgradient existence is essentially used. However, in many applications it is crucial to have an efficient solution for *non-convex optimization* problems [MB13, SFLS14, TGL13]. In particular, resource allocation problems with non-elastic traffic are studied in [TGL13]. Such applications cannot be modeled by means of concave utility functions that render a problem, a non-convex optimization, for which more sophisticated solution techniques are needed. In [MB13], the authors introduce a distributed algorithm for non-convex constrained optimization based on a first order numerical method. However, convergence to local minima is guaranteed only under the assumption that the communication topology is time-invariant, the initial values of the agents are close enough to a local minimum, and a sufficiently small step-size is used. An approximate dual subgradient algorithm over time-varying network topologies is proposed in [ZM13a]. This algorithm converges to a pair of primal-dual solutions for the approximate problem, given the Slater's condition for a constrained optimization problem, and under the assumption that the optimal solution set of the dual limit is singleton. Another work [BJ13] deals with the distributed projected stochastic gradient algorithm applicable to non-convex optimization. The authors demonstrate convergence of the proposed procedure to the set of Karush–Kuhn–Tucker points that represent only a necessary condition for a point to be a local minimum. Moreover, the work [BJ13] requires the double-stochasticity of the communication matrix on average, which may also restrict the range of potential applications. To the best of the author's knowledge, no method has been proposed so far that can guarantee convergence to a local minimum in a general case of smooth non-convex functions under a simple broadcast protocol. This chapter rectifies this issue, relaxing the assumption on the convexity and applying the push-sum algorithm to a more general non-convex case.

The distributed and recursive manner of the push-sum algorithm's performance gives a motivation to apply the procedure to learning local potential function maximizers in potential games with no memory. Note that in the context of game-theoretic applications to optimization in multi-agent systems, which is considered in this work (see Sect. 2.2), we are not interested in learning Nash equilibria in the modeled potential games. Indeed, Nash equilibria in a potential game are not generally contained in the set of local maxima of the corresponding potential function (see Fig. 2.2), which coincides with the system's global objective function. That is why the focus is rather on learning algorithms leading to optima (local or global ones) of the potential function.

As the push-sum algorithm is based on a communication protocol, the existence of the *communication* links between neighbors belongs to the information requirement in systems under consideration. Some learning procedures using communication between agents have been recently presented in the literature. In [LM13] a potential game is modeled to handle optimization problems in multi-agent systems with available communication. However, the assumption on convexity of the objective function is crucial for the efficient performance of the algorithm. Various gossip protocols are proposed as learning algorithms in games. For example, in the work [KNS12] a gossip algorithm is efficiently applied to aggregative games on graphs. The result is extended by the authors in [SP14] to the case of general continuous action games. However, behavior of procedures in both papers is studied again only for convex settings. Empirical Centroid Fictitious Play based on communication between players is proposed in [SKX15]. It is shown that the average empirical frequency of the players' actions in this Play converges to a subset of Nash equilibria in discrete action potential games. The centroid nature of the procedure allows minimizing the memory requirement. However, each player needs to choose her best response at each step, which may be rather difficult to implement. Moreover, the communication topology is complicated in all papers mentioned above. In contrast to the approaches of that papers, the chapter presents the memoryless *communication-based* algorithm, in which agents utilize the push-sum protocol. The learning rules do not require restrictive assumptions on the communication topology. This algorithm can be applied to a general non-convex continuous action potential game and guarantees almost sure convergence of agents' actions to a local maximum of the potential function.

Another algorithm presented in this chapter is based on memoryless payoff-based information (see Sect. 2.2.2) and is called *payoff-based learning*. This procedure also uses the idea of the stochastic approximation Robbins–Monro procedure. In this approach agents can only observe their currently played actions and the corresponding utility values. Among payoff-based algorithms, presented in the literature, the payoff-based version of the logit dynamics [MS12], payoff-based inhomogeneous partially irrational play [GHF12], [ZM13b], adaptive Q-learning, and ε-greedy decision rule [CLRJ13] should be mentioned. However, all these algorithms require memory and can be applied only to discrete optimization problems and, thus, to discrete action games. This fact restricts the applications of the algorithms above in control, engineering, and economics. That is why the focus is on the continuous action potential games, to which payoff-based procedure needs to be applied.

The first part of this chapter proves convergence of *the push-sum distributed optimization algorithm* to a critical point of a sum of non-convex smooth functions under some general assumptions on the underlying functions and general connectivity assumptions on the underlying graph. It will be shown that a perturbation added to the procedure allows us to use the algorithm to search local optima of this sum. It means that the stochastic procedure steers every node to some local minimum of the objective function from any initial state, if the objective function has no saddle points. The analysis uses the result on stochastic recursive approximation procedures extensively studied in [NK73]. The analysis of the convergence rate for the procedure under consideration is also provided. The second part of the chapter is devoted to application of stochastic approximation technique to learning local optima in potential games. Continuous action potential games are considered. It is shown that the analogue of the push-sum algorithm in a game, that is call *communication-based algorithm*, converges almost surely to a local maximum of the potential function, given assumptions analogous to ones in analysis of the push-sum algorithm. Another application of the stochastic approximation Robbins–Monro procedure is also considered. Namely, a *payoff-based* version of the memoryless learning procedure is presented. Under some assumptions on the potential function, the algorithm is demonstrated to converge in probability to a local maximum of the potential function. Moreover, this payoff-based approach is also applicable to problems of learning in a concave potential game with actions from compact sets. By using the properties of concave functions on compact convex sets, the chapter proves convergence of agents' actions in probability to a Nash equilibrium, which also maximizes the potential function over the set of joint actions in this case.

Some results of this chapter are presented in the works [TT16, TT15, Tat16a].

4.2 Preliminaries: Iterative Markov Process

In the following we will use the results on convergence of the stochastic approximation Robbins–Monro procedure presented in [NK73]. These results are summarized in this section. Let us start by introducing some important notations. Let $\{\mathbf{X}(t)\}_t$, $t \in \mathbb{Z}^+$, be a discrete time Markov process on some state space $E \subseteq \mathbb{R}^d$. The transition function of this chain, namely $\Pr\{\mathbf{X}(t+1) \in \Gamma | \mathbf{X}(t) = \mathbf{X}\}$, is denoted by $P(t, \mathbf{X}, t+1, \Gamma)$, $\Gamma \subseteq E$.

Definition 4.2.1 The operator L defined on the set of measurable functions $V : \mathbb{Z}^+ \times E \to \mathbb{R}$, $\mathbf{X} \in E$, by

$$\begin{aligned} LV(t,\mathbf{X}) &= \int P(t,\mathbf{X},t+1,dy)[V(t+1,y) - V(t,\mathbf{X})] \\ &= E[V(t+1,\mathbf{X}(t+1)) \mid \mathbf{X}(t) = \mathbf{X}] - V(t,\mathbf{X}). \end{aligned}$$

is called a *generating operator* of a Markov process $\{\mathbf{X}(t)\}_t$.

Let B be a subset of E, $U_\epsilon(B)$ be its ϵ-neighborhood, i.e., $U_\epsilon(B) = \{\mathbf{X} : \rho(\mathbf{X}, B) < \epsilon\}$. Let $V_\epsilon(B) = E \setminus U_\epsilon(B)$ and $U_{\epsilon,R}(B) = V_\epsilon(B) \cap \{\mathbf{X} : \|\mathbf{X}\| < R\}$.

Definition 4.2.2 The function $\psi(t, \mathbf{X})$ is said to belong to *class* $\Psi(B)$, $\psi(t, \mathbf{X}) \in \Psi(B)$, if

1) $\psi(t, \mathbf{X}) \geq 0$ for all $t \in \mathbb{Z}^+$ and $\mathbf{X} \in E$,
2) for all $R > \epsilon > 0$ there exists some $Q = Q(\epsilon, R)$ such that

$$\inf_{t \geq Q, \mathbf{X} \in U_{\epsilon,R}(B)} \psi(t, \mathbf{X}) > 0.$$

Further we consider the following general recursive process $\{\mathbf{X}(t)\}_t$ taking values in $\mathbb{R}^d$ ($E = \mathbb{R}^d$):

$$\mathbf{X}(t+1) = \mathbf{X}(t) + \alpha(t+1)\boldsymbol{F}(t, \mathbf{X}(t)) + \beta(t+1)\mathbf{W}(t+1, \mathbf{X}(t), \omega), \tag{4.1}$$

where $\mathbf{X}(0) = \boldsymbol{x}_0 \in \mathbb{R}^d$, $\boldsymbol{F}(t, \mathbf{X}(t)) = \mathbf{f}(\mathbf{X}(t)) + \mathbf{q}(t, \mathbf{X}(t))$ such that $\mathbf{f} : \mathbb{R}^d \to \mathbb{R}^d$, $\mathbf{q}(t, \mathbf{X}(t)) : \mathbb{R} \times \mathbb{R}^d \to \mathbb{R}^d$, $\{\mathbf{W}(t, \boldsymbol{x}, \omega)\}_t$ is a sequence of random vectors such that

$$\Pr\{\mathbf{W}(t+1, \mathbf{X}(t), \omega)|\mathbf{X}(t), \dots, \mathbf{X}(0)\} = \Pr\{\mathbf{W}(t+1, \mathbf{X}(t), \omega)|\mathbf{X}(t)\},$$

and $\alpha(t)$, $\beta(t)$ are some positive step-size parameters. Under the assumption on the random vectors $\mathbf{W}(t)$ above, the process (4.1) is a Markov process. We use the notation $A(t, \boldsymbol{x})$ to denote the matrix with elements $\mathrm{E}W_i(t+1, \boldsymbol{x}, \omega)W_j(t+1, \boldsymbol{x}, \omega)$.

Now we quote the following theorems for the process (4.1), which are proven in [NK73] (Theorems 2.5.2 and 2.7.3, respectively).

Theorem 4.2.1 *Consider the Markov process defined by* (4.1) *and suppose that there exists a function* $V(t, \mathbf{X}) \geq 0$ *such that* $\inf_{t\geq 0} V(t, \mathbf{X}) \to \infty$ *as* $\|\mathbf{X}\| \to \infty$ *and*

$$LV(t, \mathbf{X}) \leq -\alpha(t+1)\psi(t, \mathbf{X}) + g(t)(1 + V(t, \mathbf{X})),$$

where $\psi \in \Psi(B)$ *for some set* $B \subset \mathbb{R}^d$, $g(t) > 0$, $\sum_{t=0}^{\infty} g(t) < \infty$. *Let* $\alpha(t)$ *be such that* $\alpha(t) > 0$, $\sum_{t=0}^{\infty} \alpha(t) = \infty$. *Then almost surely* $\sup_{t\geq 0} \|\mathbf{X}(t)\| = R(\omega) < \infty$.[1]

Theorem 4.2.2 *Consider the Markov process defined by* (4.1) *and suppose that* $B = \{\mathbf{X} : \mathbf{f}(\mathbf{X}) = \mathbf{0}\}$ *be a finite set. Let the set* B, *the sequence* $\gamma(t)$, *and some function* $V(t, \mathbf{X}) : \mathbb{Z}^+ \times \mathbb{R}^d \to \mathbb{R}$ *satisfy the following assumptions:*

(1) For all $t \in \mathbb{Z}^+$, $\mathbf{X} \in \mathbb{R}^d$, *we have* $V(t, \mathbf{X}) \geq 0$ *and* $\inf_{t\geq 0} V(t, \mathbf{X}) \to \infty$ *as* $\|\mathbf{X}\| \to \infty$,

(2) $LV(t, \mathbf{X}) \leq -\alpha(t+1)\psi(t, \mathbf{X}) + g(t)(1 + V(t, \mathbf{X}))$, *where* $\psi \in \Psi(B)$, $g(t) > 0$, *and* $\sum_{t=0}^{\infty} g(t) < \infty$,

(3) $\alpha(t) > 0$, $\sum_{t=0}^{\infty} \alpha(t) = \infty$, $\sum_{t=0}^{\infty} \alpha^2(t) < \infty$, *there exists* $c(t)$: $\|\mathbf{q}(t, \mathbf{X}(t))\| \leq c(t)$ *a.s. and* $\sum_{t=0}^{\infty} \alpha(t+1)c(t) < \infty$,

(4) $\sup_{t\geq 0, \|\mathbf{X}\|\leq r} \|\boldsymbol{F}(t, \mathbf{X})\| = k(r) < \infty$, *and*

(5) $\mathrm{E}\mathbf{W}(t) = \mathbf{0}$, $\sum_{t=0}^{\infty} \mathrm{E}\beta^2(t)\|\mathbf{W}(t)\|^2 < \infty$.

Then $\Pr\{\lim_{t\to\infty} \mathbf{X}(t) = \mathbf{X}_0 \in B\} = 1$, *i.e., the process* (4.1) *converges almost surely to a point from* B.

Thus, if $\mathbf{f}(\boldsymbol{x})$ is the gradient of some objective function $\Phi(\boldsymbol{x})$: $\mathbb{R}^d \to \mathbb{R}$, then the procedure (4.1) converges almost surely to a critical point of Φ as $t \to \infty$. However, the theorem above cannot guarantee almost sure convergence to a local maxima of

[1]Here ω is an element of the sample space of the probability space on which the process $\mathbf{X}(t)$ is defined.

the objective function Φ. To exclude convergence to critical points that are not local maxima, one can use the following theorem proven in [NK73] (Theorem 5.4.1).

Theorem 4.2.3 *Consider the Markov process* $\{\mathbf{X}(t)\}_t$ *on* $\mathbb{R}^d$ *defined by* (4.1). *Let* H' *be the set of the points* $\mathbf{x}' \in \mathbb{R}^d$ *for which there exists a symmetric positive definite matrix* $C = C(\mathbf{x}')$ *and a positive number* $\epsilon = \epsilon(\mathbf{x}')$ *such that* $(\mathbf{f}(\mathbf{x}), C(\mathbf{x} - \mathbf{x}')) \geq 0$ *for* $\mathbf{x} \in U_\epsilon(\mathbf{x}')$. *Assume that for any* $\mathbf{x}' \in H'$ *there exist positive constants* $\delta = \delta(\mathbf{x}')$ *and* $K = K(\mathbf{x}')$ *such that*

(1) $\|\mathbf{f}(\mathbf{x})\|^2 + |\mathrm{Tr}[A(t, \mathbf{x}) - A(t, \mathbf{x}')]| \leq K\|\mathbf{x} - \mathbf{x}'\|$ *for any* $\mathbf{x} : \|\mathbf{x} - \mathbf{x}'\| < \delta$,
(2) $\sup_{t \geq 0, \|x - x'\| < \delta} \mathrm{E}\|\mathbf{W}(t, \mathbf{x}, \omega)\|^4 < \infty$.

Moreover,

(3) $\sum_{t=0}^{\infty} \beta^2(t) < \infty$, $\sum_{t=0}^{\infty} \left(\frac{\beta(t)}{\sqrt{\sum_{k=t+1}^{\infty} \beta^2(k)}} \right)^3 < \infty$,
(4) $\sum_{t=0}^{\infty} \frac{\alpha(t) q(t)}{\sqrt{\sum_{k=t+1}^{\infty} \beta^2(k)}} < \infty$, $q(t) = \sup_{\mathbf{x} \in \mathbb{R}^d} \|\mathbf{q}(t, \mathbf{x})\|$.

Then $\Pr\{\lim_{t \to \infty} \mathbf{X}(t) \in H'\} = 0$ *irrespectively of the initial state* x_0.

4.3 Push-Sum Algorithm in Non-convex Distributed Optimization

4.3.1 Problem Formulation: Push-Sum Algorithm and Assumptions

Consider a network of n agents. At each time t, node i can only communicate to its out-neighbors in some directed graph $G(t)$, where the graph $G(t)$ has the vertex set $[n]$ and the edge set $E(t)$. The following standard definition for the sequence $G(t)$ is introduced.

Definition 4.3.1 We say that a sequence of graphs $\{G(t)\}$ is *S-strongly connected*, if for any time $t \geq 0$, the graph

$$G(t : t + S) = ([n], E(t) \cup E(t + 1) \cup \cdots \cup E(t + S - 1)),$$

is strongly connected. In other words, the union of the graphs over every S time intervals is strongly connected.

The assumption on the S-strongly connected sequence of communication graphs has been used in many prior works [NO09, NO15, TBA+86] to ensure enough mixing of information among the agents.

$N_i^{\mathrm{in}}(t)$ and $N_i^{\mathrm{out}}(t)$ is used to denote the in- and out-neighborhoods of node i at time t. Each node i is always considered to be an in- and out-neighbor of itself. We use $d_i(t)$ to denote the out-degree of node i, and we assume that every node i knows its out-degree at every time t.

The goal of the agents is to solve distributively the following minimization problem:

$$\min_{\mathbf{z}\in\mathbb{R}^d} F(\mathbf{z}) = \sum_{i=1}^{n} F_i(\mathbf{z}). \tag{4.2}$$

The essential assumption in many distributed optimization problems is that the function $F_i : \mathbb{R}^d \to \mathbb{R}$ is only available to agent i.

In this part the following assumption on the gradients of F_i is made:

Assumption 4.3.1 *Each function $F_i(\mathbf{z})$, $i \in [n]$, has the gradient $\mathbf{f}_i(\mathbf{z}) = \nabla F_i(\mathbf{z})$ such that the norm of $\mathbf{f}_i(\mathbf{z})$ is uniformly bounded by some finite constant $\alpha \geq 0$ for each $i \in [n]$, i.e., $\|\mathbf{f}_i(\mathbf{z})\| \leq \alpha$ for all $\mathbf{z} \in \mathbb{R}^d$ and $i \in [n]$.*

Note that Assumption 4.3.1 is a standard assumption that is made in many of the previous works on subgradient methods (including [NO15]). We will further assume that a solution of (4.2) exists, and the set of critical points of the objective function $F(\mathbf{z})$, i.e., the set of points $\mathbf{z}$ such that $\nabla F(\mathbf{z}) = \mathbf{0}$, is a union of finitely many connected components.

Next, let us discuss the general push-sum algorithm initially proposed in [KDG03], applied in [TLR12a] to distributed optimization of convex functions, and analyzed in [NO15]. For this algorithm, there is a network of n agents introduced in Sect. 4.6.1. According to the general push-sum protocol, at every moment of time $t \in \mathbb{Z}^+$ each node i maintains vector variables $\mathbf{z}_i(t)$, $\mathbf{x}_i(t)$, $\mathbf{w}_i(t) \in \mathbb{R}^d$, as well as a scalar variable $y_i(t)$ with $y_i(0) = 1$ for all $i \in [n]$. These quantities are updated according to the following rules:

$$\begin{aligned}
\mathbf{w}_i(t+1) &= \sum_{j\in N_i^{\text{in}}(t)} \frac{\mathbf{x}_j(t)}{d_j(t)},\\
y_i(t+1) &= \sum_{j\in N_i^{\text{in}}(t)} \frac{y_j(t)}{d_j(t)},\\
\mathbf{z}_i(t+1) &= \frac{\mathbf{w}_i(t+1)}{y_i(t+1)},\\
\mathbf{x}_i(t+1) &= \mathbf{w}_i(t+1) + \mathbf{e}_i(t+1),
\end{aligned} \tag{4.3}$$

where $\mathbf{e}_i(t)$ is some d-dimensional, possibly random, perturbation at time t.

Theorem 4.3.1 ([NO15]) *Consider the sequences $\{\mathbf{z}_i(t)\}_t$, $i \in [n]$, generated by the algorithm (4.3). Assume that the graph sequence $\{G(t)\}$ is S-strongly connected. Then for some constants δ and λ satisfying $\delta \geq \frac{1}{n^{nS}}$ and $\lambda \leq \left(1 - \frac{1}{n^{nS}}\right)^{1/S}$ for all $i \in [n]$ we have*

$$\|\mathbf{z}_i(t+1) - \bar{\mathbf{x}}(t)\| \leq \frac{8}{\delta}\left(\lambda^t \sum_{j=1}^{n} \|\mathbf{x}_j(0)\|_1 + \sum_{s=1}^{t} \lambda^{t-s} \sum_{j=1}^{n} \|\mathbf{e}_j(s)\|_1\right).$$

Moreover, if $\{a(t)\}$ *is a non-increasing positive scalar sequence with* $\sum_{t=1}^{\infty} a(t)\|\mathbf{e}_i(t)\|_1 < \infty$ *a.s. for all* $i \in [n]$, *then*

$$\sum_{t=0}^{\infty} a(t+1)\,\|\mathbf{z}_i(t+1) - \bar{\mathbf{x}}(t)\|$$

$$\leq \sum_{t=0}^{\infty} \frac{8a(t+1)}{\delta}\left(\lambda^t \sum_{j=1}^{n} \|\mathbf{x}_j(0)\|_1 + \sum_{s=1}^{t} \lambda^{t-s} \sum_{j=1}^{n} \|\mathbf{e}_j(s)\|_1\right)$$

$$< \infty \text{ almost surely for all } i,$$

where $\|\cdot\|_1$ *is the* l^1*-norm in* $\mathbb{R}^d$.

Note that the above theorem implies that, if $\lim_{t\to\infty} \|\mathbf{e}(t)\|_1 = 0$ a.s., then

$$\lim_{t\to\infty} \|\mathbf{z}_i(t+1) - \bar{\mathbf{x}}(t)\| = 0 \quad \text{a.s. for all } i.$$

In other words, under some assumptions on the perturbations $\mathbf{e}_i(t)$ one can guarantee that in the push-sum algorithm all $\mathbf{z}_i(t)$ track the average state $\bar{\mathbf{x}}(t)$.

Similar to [NO15], let us adopt the push-sum algorithm (4.3) to the distributed optimization problem (4.2) by letting

$$\mathbf{e}_i(t+1) = -a(t+1)\mathbf{f}_i(\mathbf{z}_i(t+1)) \tag{4.4}$$

or

$$\mathbf{e}_i(t+1) \;= -a(t+1)(\mathbf{f}_i(\mathbf{z}_i(t+1)) + \mathbf{W}_i(t+1)), \tag{4.5}$$

where $\mathbf{W}_i(t+1)$ is a random vector whose entries are independent random variables with zero mean and bounded variance.

Under Assumption 4.3.1 and given that $\lim_{t\to\infty} a(t) = 0$, in both cases

$$\lim_{t\to\infty} \|\mathbf{e}_i(t)\|_1 = 0 \quad \text{a.s. for all } i.$$

Thus, in long run, the nodes' variables $\mathbf{z}_i(t+1)$ will track the average state $\bar{\mathbf{x}}(t) = \frac{1}{n}\sum_{j=1}^{n} \mathbf{x}_j(t)$ (see Theorem 4.3.1). Hence, one can expect that the long run behavior of the iterations in (4.3) with $\mathbf{e}_i(t+1)$ from (4.4) is the same as the behavior of the gradient descent iteration:

$$\bar{\mathbf{x}}(t+1) = \bar{\mathbf{x}}(t) - a(t+1)\frac{1}{n}\sum_{i=1}^{n} \mathbf{f}_i(\bar{\mathbf{x}}(t)),$$

whereas the long run behavior of the iterations in (4.3) with $\mathbf{e}_i(t+1)$ from (4.5) is equivalent to the behavior of the Robbins–Monro iteration [RM51]:

$$\bar{\mathbf{x}}(t+1) = \bar{\mathbf{x}}(t) - a(t+1)\left(\frac{1}{n}\sum_{i=1}^{n}\mathbf{f}_i(\bar{\mathbf{x}}(t)) + \bar{\mathbf{W}}(t+1)\right).$$

4.3.2 Convergence to Critical Points

First, let us recall the distributed optimization algorithm using the push-sum algorithm. At every moment of time $t \in \mathbb{Z}^+$ each node i maintains vector variables $\mathbf{z}_i(t)$, $\mathbf{x}_i(t)$, $w_i(t) \in \mathbb{R}^d$, as well as a scalar variable $y_i(t)$: $y_i(0) = 1$ for all $i \in [n]$. These quantities are updated according to (4.3) and

$$\mathbf{x}_i(t+1) = w_i(t+1) - a(t+1)\mathbf{f}_i(\mathbf{z}_i(t+1)). \tag{4.6}$$

The process above is a special case of the *perturbed push-sum* algorithm, whose background and properties are discussed in Sect. 4.3.1. According to (4.6), the average state $\bar{\mathbf{x}}(t)$ follows the dynamics

$$\bar{\mathbf{x}}(t+1) = \bar{\mathbf{x}}(t) - a(t+1)\frac{1}{n}\sum_{i=1}^{n}\mathbf{f}_i(\mathbf{z}_i(t+1)), \tag{4.7}$$

which can be rewritten as

$$\bar{\mathbf{x}}(t+1) = \bar{\mathbf{x}}(t) - a(t+1)\left(\mathbf{f}(\bar{\mathbf{x}}(t)) + \left[\frac{1}{n}\sum_{i=1}^{n}\mathbf{f}_i(\mathbf{z}_i(t+1)) - \mathbf{f}(\bar{\mathbf{x}}(t))\right]\right), \tag{4.8}$$

where $\mathbf{f}(\mathbf{z}) = \frac{1}{n}\sum_{i=1}^{n}\mathbf{f}_i(\mathbf{z}) = \frac{1}{n}\nabla F(\mathbf{z})$. If we denote $\frac{1}{n}\sum_{i=1}^{n}\mathbf{f}_i(\mathbf{z}_i(t+1)) - \mathbf{f}(\bar{\mathbf{x}}(t))$ by $\mathbf{q}(t, \bar{\mathbf{x}}(t))$, the process (4.8) becomes a deterministic version of the process (4.1) ($\mathbf{W}(t) \equiv \mathbf{0}$ for any t). It is shown in [NO15] that if functions $F_i(\mathbf{z}_i)$ are convex functions satisfying Assumption 4.3.1, and there exists a solution of (4.2), then the process (4.7) converges to an optimizer of the function F, given an appropriate choice of step-size sequence $a(t)$.

Let the gradients $\mathbf{f}_i(\mathbf{x})$, $i \in [n]$, satisfy the following assumption:

Assumption 4.3.2 *For each $i \in [d]$, $\mathbf{f}_i$ is Lipschitz continuous on $\mathbb{R}^d$, i.e., there exists a constant $l_i \geq 0$ such that $\|\mathbf{f}_i(\mathbf{x}_1) - \mathbf{f}_i(\mathbf{x}_2)\| \leq l_i\|\mathbf{x}_1 - \mathbf{x}_2\|$ for any $\mathbf{x}_1, \mathbf{x}_2 \in \mathbb{R}^d$.*

Further the following assumption on the behavior of the objective function $F(\mathbf{z})$, when $\|\mathbf{z}\| \to \infty$, is needed.

Assumption 4.3.3 *$F(\mathbf{z})$ is coercive, i.e.,* $\lim_{\|\mathbf{z}\|\to\infty} F(\mathbf{z}) \to \infty$.[2]

Let us denote the set of critical points of F by B, i.e.,

$$B = \{\mathbf{z} \in \mathbb{R}^d : \mathbf{f}(\mathbf{z}) = \mathbf{0}\}.$$

B'' is used to represent its refinement to the set of local minima, i.e.,

$$B'' = \{\mathbf{z} \in B : \text{ there exists } \delta : \mathbf{f}(\mathbf{z}) \le \mathbf{f}(\mathbf{z}'), \text{ for any } \mathbf{z}' : \|\mathbf{z}' - \mathbf{z}\| < \delta\},$$

and represent the rest of the critical points by $B' = B \setminus B''$.

Now the main result of this subsection can be formulated.

Theorem 4.3.2 *Let the functions F and $\mathbf{f}_i$, $i \in [n]$, satisfy Assumptions 4.3.1–4.3.3. Let $\{a(t)\}$ be a positive and non-increasing step-size sequence such that $\sum_{t=0}^{\infty} a(t) = \infty$ and $\sum_{t=0}^{\infty} a^2(t) < \infty$. Then the average state $\bar{\mathbf{x}}(t)$ and $\mathbf{z}_i(t)$ (see (4.8)) for the dynamics (4.6) with S-strongly connected graph sequence $\{G(t)\}$ converge either to a point from the set B or to the boundary of one of its connected components.*

Before proving this theorem, let us note that under Assumption 4.3.2 each gradient function $\mathbf{f}_i(\mathbf{x})$, $i \in [n]$, is a Lipschitz function with some constant l_i. This allows us to formulate the following lemma which will be needed for proof of Theorem 4.3.2.

Lemma 4.3.1 *Let $\{a(t)\}$ be a non-increasing sequence such that $\sum_{t=0}^{\infty} a^2(t) < \infty$. Then, there exists $c(t)$ such that the following holds for the process (4.8), given Assumptions 4.3.1, 4.3.2 and a S-strongly connected graph sequence $\{G(t)\}$:*

$$\|\mathbf{q}(t, \bar{\mathbf{x}}(t))\| = \left\| \frac{1}{n} \sum_{i=1}^{n} \mathbf{f}_i(\mathbf{z}_i(t+1)) - \mathbf{f}(\bar{\mathbf{x}}(t)) \right\| \le c(t),$$

and

$$\sum_{t=0}^{\infty} a(t+1)c(t) < \infty.$$

[2]Since we also require the boundedness of ∇F (Assumption 4.3.1), the function $F(\mathbf{z})$ is assumed to increase not faster than a linear function as $\mathbf{z} \to \infty$.

Proof Since the functions $\mathbf{f}_i$ are Lipschitz (Assumption 4.3.2) and taking into account Theorem 4.3.1, we have that

$$\left\| \sum_{i=1}^{n} \mathbf{f}_i(\mathbf{z}_i(t+1)) - \mathbf{f}(\bar{\mathbf{x}}(t)) \right\| \leq \sum_{i=1}^{n} \|\mathbf{f}_i(\mathbf{z}_i(t+1)) - \mathbf{f}_i(\bar{\mathbf{x}}(t))\|$$
$$\leq \sum_{i=1}^{n} l_i \|\mathbf{z}_i(t+1) - \bar{\mathbf{x}}(t)\|$$
$$\leq \frac{8\ln}{\delta} \left(\lambda^t \sum_{j=1}^{n} \|\mathbf{x}_j(0)\|_1 + \sum_{s=1}^{t} \lambda^{t-s} \sum_{j=1}^{n} \|\mathbf{e}_j(s)\|_1 \right)$$

for some positive δ and λ and where $l = \max_{i\in[n]} l_i$, $\|\mathbf{e}_i(t)\| = a(t)\|\mathbf{f}_i(\mathbf{z}_i(t))\|$.
Let

$$c(t) = \frac{8l}{\delta} \left(\lambda^t \sum_{j=1}^{n} \|\mathbf{x}_j(0)\|_1 + \sum_{s=1}^{t} \lambda^{t-s} \sum_{j=1}^{n} \|\mathbf{e}_j(s)\|_1 \right).$$

Then, $\|\mathbf{q}(t, \bar{\mathbf{x}}(t))\| \leq c(t)$. Taking into account Assumption 4.3.1, we get

$$\sum_{t=1}^{\infty} a(t)\|\mathbf{e}_i(t)\| = \sum_{t=0}^{\infty} a^2(t)\|\mathbf{f}_i(\mathbf{z}_i(t))\| \leq \alpha \sum_{t=0}^{\infty} a^2(t) < \infty,$$

Hence, according to Theorem 4.3.1,

$$\sum_{t=0}^{\infty} a(t+1)c(t) < \infty.$$

□

Proof of Theorem 4.3.2 To prove this statement we will use the general result formulated in Theorem 4.2.2. We emphasize one more time that the process (4.8) under consideration represents a special deterministic case of the recursive procedure (4.1), where $\alpha(t) = \beta(t) = -a(t)$.[3] Note that, according to the choice of $a(t)$, Assumption 4.3.1, and Lemma 4.3.1, conditions (4.2.2)–(4.2.2) of Theorem 4.2.2 hold.[4] Thus, it suffices to show that there exists a sample function $V(t, \mathbf{x})$ of the process (4.8) satisfying the conditions 4.2.2 and 4.2.2 in Theorem 4.2.2. A natural choice for such a function is the (time-invariant) function $V(t, \mathbf{x}) = V(\mathbf{x}) = F(\mathbf{x}) + C$, where C is a constant chosen to guarantee the positiveness of $V(\mathbf{x})$ over the

[3]Note that in this part we are interested in minimization problems.

[4]Note that in the deterministic case under consideration condition (4.2.2) holds automatically.

whole $\mathbb{R}^d$. Note that such constant always exists because of the continuity of F and Assumption 4.3.3. Then, $V(\mathbf{x})$ is nonnegative and $V(\mathbf{x}) \to \infty$ as $\|\mathbf{x}\| \to \infty$. Let $LV(\mathbf{x}(t)) := LV(t, \mathbf{x})$. Now we show that the function $V(\mathbf{x})$ satisfies the condition 4.2.2 in Theorem 4.2.2. Using the notation $\tilde{\mathbf{f}}(t, \bar{\mathbf{x}}(t)) = \mathbf{f}(\bar{\mathbf{x}}(t)) + \mathbf{q}(t, \bar{\mathbf{x}}(t))$, $\mathbf{q}(t, \bar{\mathbf{x}}(t)) = \frac{1}{n}\sum_{i=1}^{n} \mathbf{f}_i(\mathbf{z}_i(t+1)) - \mathbf{f}(\bar{\mathbf{x}}(t))$ and by the Mean-value Theorem:

$$\begin{aligned}
LV(\bar{\mathbf{x}}(t)) &= V(\bar{\mathbf{x}}(t+1)) - V(\bar{\mathbf{x}}(t)) = V(\bar{\mathbf{x}}(t) - a(t+1)\tilde{\mathbf{f}}(t, \bar{\mathbf{x}}(t))) - V(\bar{\mathbf{x}}(t)) \\
&= -a(t+1)(\nabla V(\tilde{\mathbf{x}}), \tilde{\mathbf{f}}(t, \bar{\mathbf{x}}(t))) \\
&= -a(t+1)(\nabla V(\bar{\mathbf{x}}(t)), \tilde{\mathbf{f}}(t, \bar{\mathbf{x}}(t))) - a(t+1)(\nabla V(\tilde{\mathbf{x}}) - \nabla V(\bar{\mathbf{x}}(t)), \tilde{\mathbf{f}}(t, \bar{\mathbf{x}}(t))),
\end{aligned}$$

where $\tilde{\mathbf{x}} = \bar{\mathbf{x}}(t) - \theta a(t+1)\tilde{\mathbf{f}}(t, \bar{\mathbf{x}}(t))$ for some $\theta \in (0, 1)$. Taking into account Assumption 4.3.2, we obtain that for some constant $k > 0$,

$$\|\nabla V(\tilde{\mathbf{x}}) - \nabla V(\bar{\mathbf{x}}(t))\| \le k\|\tilde{\mathbf{x}} - \bar{\mathbf{x}}(t)\| = ka(t+1)\theta\|\tilde{\mathbf{f}}(t, \bar{\mathbf{x}}(t))\|.$$

Hence, due to the Cauchy–Schwarz inequality,

$$\begin{aligned}
LV(\bar{\mathbf{x}}(t)) &\le -a(t+1)(\nabla V(\bar{\mathbf{x}}(t)), \tilde{\mathbf{f}}(t, \bar{\mathbf{x}}(t))) + k_1 a^2(t+1)\|\tilde{\mathbf{f}}(t, \bar{\mathbf{x}}(t))\|^2 \\
&\le -a(t+1)(\nabla V(\bar{\mathbf{x}}(t)), \mathbf{f}(\bar{\mathbf{x}}(t))) + k_2 a(t+1)\|\mathbf{q}(t, \bar{\mathbf{x}}(t))\| \\
&\quad + k_3 a^2(t+1)(1 + \|\mathbf{q}(t, \bar{\mathbf{x}}(t))\| + \|\mathbf{q}(t, \bar{\mathbf{x}}(t))\|^2)
\end{aligned}$$

for some $k_1, k_2, k_3 > 0$.

Recall that the function $V(\mathbf{x})$ is nonnegative. Thus, we finally obtain that

$$LV(\bar{\mathbf{x}}(t)) \le -a(t+1)(\nabla V(\bar{\mathbf{x}}(t)), \mathbf{f}(\bar{\mathbf{x}}(t))) + g(t)(1 + V(\bar{\mathbf{x}}(t))),$$

where

$$\begin{aligned}
g(t) &= k_4(a(t+1)\|\mathbf{q}(t, \bar{\mathbf{x}}(t))\| + a^2(t+1) \\
&\quad + a^2(t+1)\|\mathbf{q}(t, \bar{\mathbf{x}}(t))\| + a^2(t+1)\|\mathbf{q}(t, \bar{\mathbf{x}}(t))\|^2)
\end{aligned}$$

for some constant $k_4 > 0$. Lemma 4.3.1 and the choice of $a(t)$ imply that $g(t) > 0$ and $\sum_{t=0}^{\infty} g(t) < \infty$. Hence,

$$LV(\bar{\mathbf{x}}(t)) \le -a(t+1)\phi(t, \bar{\mathbf{x}}(t)) + g(t)(1 + V(\bar{\mathbf{x}}(t))),$$

where, according to the choice of $V(\mathbf{x})$,

$$\phi(t, \bar{\mathbf{x}}(t)) = (\nabla V(\bar{\mathbf{x}}(t)), \mathbf{f}(\bar{\mathbf{x}}(t))) = \|\mathbf{f}(\bar{\mathbf{x}}(t))\|^2 \in \Phi(B).$$

Thus, all conditions of Theorem 4.2.2 hold and we conclude that either $\lim_{t\to\infty} \bar{\mathbf{x}}(t) = \mathbf{z}_0 \in B$ or $\bar{\mathbf{x}}(t)$ converges to the boundary of a connected component of the set B. By Theorem 4.3.1, $\lim_{t\to\infty} \|\mathbf{z}_i(t) - \bar{\mathbf{x}}(t)\| = 0$ for all $i \in [n]$ and, hence, the result follows. □

Unfortunately (and naturally), there exists no function $V(t, \mathbf{x})$ for which

$$\phi(t, \bar{\mathbf{x}}(t)) = (\nabla V(\bar{\mathbf{x}}(t)), \mathbf{f}(\bar{\mathbf{x}}(t))) \in \Phi(B''),$$

where B'' represents the set of local minima of the function $F(\mathbf{z})$. Thus, the deterministic process (4.8) is unable to distinguish between local minima, saddle points, and local maxima and guarantees only the convergence to a zero point of the gradient ∇F. Further, a solution on how to rectify this issue will be provided.

4.3.3 Perturbed Procedure: Convergence to Local Minima

This subsection overcomes the weakness mentioned at the end of the previous subsection by modifying the deterministic process (4.8). Some noise is added to the iterative process that will render it as a Markov chain in $\mathbb{R}^d$. Here, the idea is that the non-local minima points have unstable directions (with respect to the gradient descent dynamics) that can be explored using the additional noise and hence, the algorithm will never get stuck at the non-local minima critical points. We will show that this noise together with an appropriate choice of the step-size sequence $a(t)$ allows the algorithm to converge to a local optimum and not to become stuck in a suboptimal critical point. Let us consider the following variation of (4.6):

$$\begin{aligned}
w_i(t+1) &= \sum_{j\in N_i^{\text{in}}(t)} \frac{\mathbf{x}_j(t)}{d_j(t)}, \\
y_i(t+1) &= \sum_{j\in N_i^{\text{in}}(t)} \frac{y_j(t)}{d_j(t)}, \\
\mathbf{z}_i(t+1) &= \frac{w_i(t+1)}{y_i(t+1)}, \\
\mathbf{x}_i(t+1) &= w_i(t+1) - a(t+1)\mathbf{f}_i(\mathbf{z}_i(t+1)) - a(t+1)\mathbf{W}_i(t+1),
\end{aligned} \tag{4.9}$$

where the functions $\mathbf{f}_i$ are the subgradient functions and $\{\mathbf{W}_i(t)\}_t$ is the sequence of independently and identically distributed (i.i.d.) random vectors taking values in $\mathbb{R}^d$ and satisfying the following assumption:

Assumption 4.3.4 *The entries $W_i^k(t)$, $W_i^j(t)$ of each $\mathbf{W}_i(t)$, $i \in [n]$, are independent,* $\mathrm{E}(W_i^k(t)) = 0$, *and* $\mathrm{Var}(W_i^k(t)) = 1$ *for all $t \in \mathbb{Z}^+$ and $k, j = 1, \ldots, d$.*

The procedure above implies the following update for the average state:

$$\bar{\mathbf{x}}(t+1) = \bar{\mathbf{x}}(t) - a(t+1)\left(\mathbf{f}(\bar{\mathbf{x}}(t)) + \left[\frac{1}{n}\sum_{i=1}^{n}\mathbf{f}_i(\mathbf{z}_i(t+1)) - \mathbf{f}(\bar{\mathbf{x}}(t))\right]\right)$$
$$- a(t+1)\bar{\mathbf{W}}(t+1). \tag{4.10}$$

Under Assumption 4.3.4, the process (4.10) is a Markov chain on the space $\mathbb{R}^d$ that is a special case of the process (4.1).

Theorem 4.3.3 *Let Assumptions 4.3.1–4.3.4 hold. Let $\{a(t)\}$ be a positive and non-increasing step-size sequence with $\sum_{t=0}^{\infty} a(t) = \infty$, $\sum_{t=0}^{\infty} a^2(t) < \infty$. Then the average state $\bar{\mathbf{x}}(t)$ and $\mathbf{z}_i(t)$ defined by* (4.9) *and* (4.10) *with S-strongly connected graph sequence $\{G(t)\}$ converge either to a point from the set B or to the boundary of one of its connected components.*

Proof Assumptions 4.3.1, 4.3.2, and 4.3.4 allow us to use Theorem 4.3.1 to conclude that all $\mathbf{z}_i(t+1)$ in (4.9) converge to the average state $\bar{\mathbf{x}}(t)$ almost surely, if $a(t) \to 0$ as $t \to \infty$. Moreover, if $\sum_{t=0}^{\infty} a^2(t) < \infty$, Lemma 4.3.1 holds for the process (4.10), where $\|\mathbf{q}(t, \bar{\mathbf{x}}(t))\| \le c'(t)$ a.s.,

$$c'(t) = M\left(\lambda^t + \sum_{s=1}^{t}\lambda^{t-s}a(s)\right) \tag{4.11}$$

for some positive M, and $\sum_{t=0}^{\infty} a(t+1)c'(t) < \infty$. Now, considering $V(\mathbf{x}) = F(\mathbf{x}) + C$, where C is such that $V(\mathbf{x}) > 0$ for all $\mathbf{x}$, and noticing that $LV(\bar{\mathbf{x}}(t)) = \mathrm{E}[V(\{\bar{\mathbf{x}}(t+1)|\bar{\mathbf{x}}(t) = \mathbf{x}\}) - V(\mathbf{x})]$, we can use the Mean-value Theorem for the term under the expectation (as in the proof of Theorem 4.3.2), and verify that all conditions of Theorem 4.2.2 hold for the process (4.10), given that $\alpha(t) = \beta(t) = -a(t)$, Assumption 4.3.3 holds and the step-size $\{a(t)\}$ is appropriately chosen. □

The above result is an analogue of Theorem 4.3.2 adapted for (4.9) and (4.10). It is clear that Theorem 4.3.3 does not guarantee the convergence of the process (4.10) to some local minimum of the objective function F. However, we will prove further the following theorem claiming that the process (4.10) cannot converge to a critical point that is not some local minimum of the objective function $F(\mathbf{z})$, given an appropriate step-size sequence $a(t)$, Assumptions 4.3.1–4.3.4, and

Assumption 4.3.5 *For any point $\mathbf{z}' \in B'$ that is not a local minimum of F there exists a symmetric positive definite matrix $C(\mathbf{z}')$ such that $(\mathbf{f}(\mathbf{z}), C(\mathbf{z}')(\mathbf{z} - \mathbf{z}')) \le 0$ for any $\mathbf{z} \in U(\mathbf{z}')$, where $U(\mathbf{z}')$ is some open neighborhood of $\mathbf{z}'$.*

Note that if the second derivatives of the function F exist, the assumption above holds for any critical point of F that is a local maximum. Indeed, in this case

$$\mathbf{f}(\mathbf{z}) = H\{F(\mathbf{z}')\}(\mathbf{z} - \mathbf{z}') + \delta(\|\mathbf{z} - \mathbf{z}'\|), \tag{4.12}$$

where $\delta(\|\mathbf{z}-\mathbf{z}'\|) = o(1)$ as $\mathbf{z} \to \mathbf{z}'$ and $H\{F(\cdot)\}$ is the Hessian matrix of F at the corresponding point. As $-H\{F(\mathbf{z}')\}$ is positive definite, there exists some open neighborhood $U(\mathbf{z}')$ of $\mathbf{z}'$ such that $(H\{F(\mathbf{z}')\}(\mathbf{z}-\mathbf{z}'), \mathbf{z}-\mathbf{z}') < 0$ for any $\mathbf{z} \in U(\mathbf{z}')$, $\mathbf{z} \neq \mathbf{z}'$. Hence, according to (4.12), we conclude that $(\mathbf{f}(\mathbf{z}), (\mathbf{z}-\mathbf{z}')) \leq 0$ for any $\mathbf{z} \in U(\mathbf{z}')$. However, note that Assumption 4.3.5 does not require existing of second derivatives. An example in this case would be the function $F(z)$, $z \in \mathbb{R}$ that behaves in a neighborhood of its local maximum $z = 0$ as follows:

$$F(z) = \begin{cases} -z^2, & \text{if } z \leq 0, \\ -3z^2, & \text{if } z > 0. \end{cases}$$

Obviously, the function above is not twice differentiable at $z = 0$, but $\nabla F(z)z \leq 0$ for any $z \in \mathbb{R}$.

Theorem 4.3.4 *Let the objective function $F(\mathbf{z})$ and gradients $\mathbf{f}_i(\mathbf{z})$, $i \in [n]$, in the distributed optimization problem* (4.2) *satisfy Assumptions 4.3.1–4.3.3, and 4.3.5. Let the sequence of the random i.i.d. vectors $\{\mathbf{W}_i(t)\}_t$, $i \in [n]$, satisfy Assumption 4.3.4. Let $\{a(t)\}$ be a positive and non-increasing step-size sequence such that $a(t) = O\left(\frac{1}{t^\nu}\right)$, where $\frac{1}{2} < \nu \leq 1$. Then the average state vector $\bar{\mathbf{x}}(t)$ (defined by* (4.10)*) and states $\mathbf{z}_i(t)$ for the distributed optimization problem* (4.9) *with S-strongly connected graph sequence $\{G(t)\}$ converge almost surely to a point from the set $B'' = B \setminus B'$ of the local minima of the function $F(\mathbf{z})$ or to the boundary of one of its connected components, for any initial states $\{\mathbf{x}_i(0)\}_{i\in[n]}$.*

Remark 4.3.1 If in Assumption 4.3.5 we assume that B' is defined as a set of points $\mathbf{z}' \in \mathbb{R}^d$ for which there exists a symmetric positive definite matrix $C(\mathbf{z}')$ such that $(\mathbf{f}(\mathbf{z}), C(\mathbf{z}')(\mathbf{z}-\mathbf{z}')) \leq 0$ for any $\mathbf{z} \in U(\mathbf{z}')$, then Theorem 4.3.4 claims the almost sure convergence of the process in (4.10) to a point from the set of critical point defined by $B'' = B \setminus B'$ or to the boundary of one of its connected components.

Proof of Theorem 4.3.4 We first notice that, under the proposed choice of $a(t)$,

$$\sum_{t=0}^{\infty} a(t) = \infty \text{ and } \sum_{t=0}^{\infty} a^2(t) < \infty.$$

Hence, we can use Theorem 4.3.3 to conclude that the process (4.10) converges to the set B represented by critical points of the function $F(\mathbf{z})$. Recall that the set of points that are not local minima is denoted by B' and $B'' = B \setminus B'$ is the set of local minima of the function $F(\mathbf{z})$. Further we apply the result formulated in Theorem 4.2.3 to the process (4.10), where $\alpha(t) = \beta(t) = -a(t)$, and take into account that we are looking for a solution for a minimization problem. Let us notice that, according to Assumption 4.3.5, for any $\mathbf{z}' \in B'$ there exist some neighborhood $U_\epsilon(\mathbf{z}')$ and some symmetric positive definite matrix $C = C(\mathbf{z}')$ such that $(\mathbf{f}(\mathbf{z}), C(\mathbf{z}-\mathbf{z}')) \leq 0$ for $\mathbf{z} \in U_\epsilon(\mathbf{z}')$. Moreover, since $B' \subseteq B = \{\mathbf{z} : \mathbf{f}(\mathbf{z}) = \mathbf{0}\}$

and due to Assumptions 4.3.2 and 4.3.4, we can conclude that condition (4.2.3) of Theorem 4.2.3 holds for B' and $\mathbf{f}$.

It is straightforward to verify that the sequence $a(t) = O\left(\frac{1}{t^\nu}\right)$ satisfies condition (3) of Theorem 4.2.3. Next, recall (see (4.11)) that

$$\|\mathbf{q}(t, \bar{\mathbf{x}}(t))\| \leq \frac{1}{n}\sum_{i=1}^{n} l_i \|\mathbf{z}_i(t+1) - \bar{\mathbf{x}}(t)\| \leq M\left(\lambda^t + \sum_{s=1}^{t} \lambda^{t-s} a(s)\right)$$

almost surely for some positive constant M. Hence,

$$q(t) = \sup_{\mathbf{x}\in\mathbb{R}^d} \|\mathbf{q}(t, \mathbf{x})\| \leq M\left(\lambda^t + \sum_{s=1}^{t} \lambda^{t-s} a(s)\right).$$

Taking into account this fact and the fact that $\frac{a(t)}{\sqrt{\sum_{k=t+1}^{\infty} a^2(k)}} = O(\frac{1}{\sqrt{t}})$, we have

$$\sum_{t=0}^{\infty} \frac{a(t)q(t)}{\sqrt{\sum_{k=t+1}^{\infty} a^2(k)}} \leq \sum_{t=0}^{\infty} O\left(\frac{1}{\sqrt{t}}\right)\left(\lambda^t + \sum_{s=1}^{t} \lambda^{t-s} O\left(\frac{1}{s^\nu}\right)\right) < \infty.$$

The last inequality is due to the following considerations.

(1) $$\sum_{t=0}^{\infty} O\left(\frac{1}{\sqrt{t}}\right)\lambda^t \leq \sum_{t=0}^{\infty} O(\lambda^t) < \infty, \text{ since } \lambda \in (0, 1).$$

(2) $$\sum_{t=0}^{\infty} O\left(\frac{1}{\sqrt{t}}\right)\left(\sum_{s=1}^{t} \lambda^{t-s} O\left(\frac{1}{s^\nu}\right)\right) \leq \sum_{t=0}^{\infty}\sum_{s=1}^{t} \lambda^{t-s} O\left(\frac{1}{s^{\nu+1/2}}\right) < \infty,$$

since, according to [RNV10], any series of the type $\sum_{k=0}^{\infty}\left(\sum_{l=0}^{k} \beta^{k-l}\gamma_l\right)$ converges, if $\gamma_k \geq 0$ for all $k \geq 0$, $\sum_{k=0}^{\infty} \gamma_k < \infty$, and $\beta \in (0, 1)$.

Thus, all conditions of Theorem 4.2.3 are fulfilled for the points from the set B'. It implies that the process (4.10) cannot converge to the points from the set B' and, thus, $\Pr\{\lim_{t\to\infty} \bar{\mathbf{x}}(t) = \mathbf{z}_0 \in B''\} = 1$. We conclude the proof by noting that, according to Theorem 4.3.1, $\lim_{t\to\infty} \|\mathbf{z}_i(t) - \bar{\mathbf{x}}(t)\| = 0$ almost surely for all $i \in [n]$ and, hence, the result follows. □

4.3.4 Convergence Rate of the Perturbed Process

This subsection presents a result on the convergence rate of the procedure (4.10) introduced above. Recall that Theorem 4.3.4 claims almost sure convergence of this process to a local minimum of the function in the distributed optimization problem (4.2), given Assumptions 4.3.1–4.3.5. To formulate the result on the convergence rate, we need the following assumption on gradients' smoothness.

Assumption 4.3.6 $\frac{\partial f^k(x)}{\partial x^l}$ *exists and is bounded for all* $k, l = 1 \dots d$, *where* f^k *is the k-th coordinate of the vector* $\mathbf{f}$.

Let us start by revisiting a well-known result in linear systems theory [Mal52].

Lemma 4.3.2 *If some matrix A is stable,*[5] *then for any symmetric positive definite matrix D, there exists a symmetric positive definite matrix C such that* $CA + A^T C = -D$. *In fact,* $C = \int_0^\infty e^{A^T\tau} D e^{A\tau} d\tau$.

Recall that the objective function $F(\mathbf{z})$ under consideration has finitely many local minima $\{\mathbf{x}_1^*, \dots, \mathbf{x}_R^*\}$. We begin by noticing that according to Assumption 4.3.6 the function $\mathbf{f}(\mathbf{x})$ admits the following representation in some neighborhood of any local minimum $\mathbf{x}_m^*$, $m = 1, \dots, R$,

$$\mathbf{f}(\mathbf{x}) = H\{F(\mathbf{x}_m^*)\}(\mathbf{x} - \mathbf{x}_m^*) + \delta_m(\|\mathbf{x} - \mathbf{x}_m^*\|),$$

where $\delta_m(\|\mathbf{x} - \mathbf{x}_m^*\|) = o(1)$ as $\mathbf{x} \to \mathbf{x}_m^*$ and $H\{F(\cdot)\}$ is the Hessian matrix of F at the corresponding point.

For the sake of notational simplicity, let the matrix $H\{F(\mathbf{x}_m^*)\}$ be denoted by H_m. Now the following lemma can be formulated.

Lemma 4.3.3 *Under Assumption 4.3.6, for any local minimum* $\mathbf{x}_m^*$, $m = 1, \dots, R$, *of the objective function F there exist a symmetric positive definite matrix* C_m *and positive constants* $\beta(m)$, $\varepsilon(m)$, *and* $a < \infty$ *(independent on m), such that for any* $\mathbf{x}$: $\|\mathbf{x} - \mathbf{x}_m^*\| < \varepsilon(m)$ *the following holds:*

$$(C_m \mathbf{f}(\mathbf{x}), \mathbf{x} - \mathbf{x}_m^*) \geq \beta(m)(C_m(\mathbf{x} - \mathbf{x}_m^*), \mathbf{x} - \mathbf{x}_m^*),$$
$$2a\beta(m) > 1.$$

Proof Since $\mathbf{x}_m^*$, for $m \in [R]$, is a local minima of F, we can conclude that H_m, $m \in [R]$, is a symmetric positive definite matrix. Hence, there exists a finite constant $a > 0$ such that the matrix $-aH_m + \frac{1}{2}I$ is stable for all $m \in [R]$, where I is the identity matrix.

Without loss of generality we assume that $\mathbf{x}_m^* = \mathbf{0}$. Let $\lambda_1(m), \dots, \lambda_d(m)$ be the eigenvalues of H_m, $\tilde{\lambda}(m) = \min_{j \in [d]} \lambda_j(m)$. Since the matrix $-aH_m + \frac{1}{2}I$ is stable, we conclude that $2a\tilde{\lambda}(m) > 1$. Moreover, $-H_m + \lambda_0(m)I$ is stable as well for any $\lambda_0(m) < \tilde{\lambda}(m)$, since the matrix $-H_m$ is stable. Hence, we can apply Lemma 4.3.2 to $-H_m + \lambda_0(m)I$ which implies that there exists a matrix $C_m = C_m(\lambda_0)$ such that

$$(C_m H_m \mathbf{x}, \mathbf{x}) \geq \lambda_0(m)(C_m \mathbf{x}, \mathbf{x}). \tag{4.13}$$

[5] A matrix is called stable (Hurwitz) if all its eigenvalues have a strictly negative real part.

Now we remind that in some neighborhood of $\mathbf{0}$

$$\mathbf{f}(\mathbf{x}) = H_m\mathbf{x} + \delta_m(\|\mathbf{x}\|), \quad \delta_m(\|\mathbf{x}\|) = o(1) \text{ as } \mathbf{x} \to \mathbf{0}.$$

Thus, taking into account (4.13), we conclude that for any $\beta(m) < \lambda_0(m)$ there exists $\varepsilon(m) > 0$ such that for any $\mathbf{x}$: $\|\mathbf{x}\| < \varepsilon(m)$

$$(C_m\mathbf{f}(\mathbf{x}), \mathbf{x}) \geq \beta(m)(C_m\mathbf{x}, \mathbf{x}).$$

As the constants $\lambda_0(m) < \tilde{\lambda}(m)$ and $\beta(m) < \lambda_0(m)$ can be chosen arbitrarily and $2a\tilde{\lambda}(m) > 1$, we conclude that

$$2a\beta(m) > 1.$$

That completes the proof. □

Now the following theorem can be formulated.

Theorem 4.3.5 *Let the objective function $F(\mathbf{z})$ have finitely many critical points, i.e., the set B be finite.*[6] *Let the objective function $F(\mathbf{z})$, gradients $\mathbf{f}_i(\mathbf{z})$, and $\mathbf{f} = \frac{1}{n}\sum_{i=1}^{n}\mathbf{f}_i$, $i \in [n]$, in the distributed optimization problem* (4.2) *satisfy Assumptions 4.3.1–4.3.3 and 4.3.5–4.3.6. Let the sequence of the random i.i.d. vectors $\{\mathbf{W}_i(t)\}_t$, $i \in [n]$, satisfy Assumption 4.3.4 and the graph sequence $\{G(t)\}$ be S-strongly connected. Then there exists a constant $\alpha > 0$ such that for any $0 < a \leq \alpha$, the average state $\bar{\mathbf{x}}(t)$ (defined by* (4.10)*) and states $\mathbf{z}_i(t)$ for the process* (4.10) *with the choice of step-size $a(t) = \frac{a}{t}$ converge to a point in B'' (the set of local minima of $F(\mathbf{z})$). Moreover, for any $\mathbf{x}^* \in B''$*

$$\mathrm{E}\{\|\bar{\mathbf{x}}(t) - \mathbf{x}^*\|^2 \mid \lim_{s\to\infty} \bar{\mathbf{x}}(s) = \mathbf{x}^*\} = O\left(\frac{1}{t}\right) \text{ as } t \to \infty.$$

Proof We proceed to show this result in two steps: we first show that a trimmed version of the dynamics converges on $O\left(\frac{1}{t}\right)$ and then, we show that the convergence rate of the trimmed dynamics and the original dynamics are the same. Without loss of generality we assume that $\mathbf{x}^* = \mathbf{0}$ and

$$\Pr\{\lim_{s\to\infty} \bar{\mathbf{x}}(s) = \mathbf{x}^* = \mathbf{0}\} > 0.$$

From Lemma 4.3.3, we know that there exist a symmetric positive definite matrix C and positive constants β, ε, and a such that $(C\mathbf{f}(\mathbf{x}), \mathbf{x}) \geq \beta(C\mathbf{x}, \mathbf{x})$, for any $\mathbf{x}$: $\|\mathbf{x}\| < \varepsilon$. Moreover, $2a\beta > 1$.

---

[6]This assumption is made to simplify notations in the proof of this theorem. Note that this theorem can be generalized to the case of finitely many connected components in the set of critical points and local minima.

Let us consider the following trimmed process $\hat{\mathbf{x}}^{\tau,\kappa}(t)$ defined by:

$$\hat{\mathbf{x}}^{\tau,\kappa}(t+1) = \hat{\mathbf{x}}^{\tau,\kappa}(t) - \frac{a}{t+1}\left(\hat{\mathbf{f}}(\hat{\mathbf{x}}^{\tau,\kappa}(t)) + \mathbf{q}(t,\bar{\mathbf{x}}(t))\right) - \frac{a}{t+1}\hat{\mathbf{W}}(t+1,\hat{\mathbf{x}}^{\tau,\kappa}(t)),$$

for $t \geq \tau$ with $\hat{\mathbf{x}}^{\tau,\kappa}(\tau) = \kappa$, where the random vector $\bar{\mathbf{x}}(t)$ is updated according to (4.10) with $a(t) = \frac{a}{t}$,

$$\hat{\mathbf{f}}(\mathbf{x}) = \begin{cases} \mathbf{f}(\mathbf{x}), & \text{if } \|\mathbf{x}\| < \varepsilon \\ \beta\mathbf{x}, & \text{if } \|\mathbf{x}\| \geq \varepsilon \end{cases},$$

$$\hat{\mathbf{W}}(t,\mathbf{x}) = \begin{cases} \mathbf{W}(t), & \text{if } \|\mathbf{x}\| < \varepsilon \\ \mathbf{0}, & \text{if } \|\mathbf{x}\| \geq \varepsilon \end{cases}.$$

Next we show the $O\left(\frac{1}{t}\right)$ rate of convergence for the trimmed process. The proof follows similar argument as of Lemma 6.2.1 of [NK73]. Obviously, $(C\hat{\mathbf{f}}(\mathbf{x}),\mathbf{x}) \geq \beta(C\mathbf{x},\mathbf{x})$, for any $\mathbf{x} \in \mathbb{R}^d$. We proceed by showing that $\mathrm{E}\|\hat{\mathbf{x}}^{\tau,\kappa}(t)\|^2 = O\left(\frac{1}{t}\right)$ as $t \to \infty$ for any $\tau \geq 0$ and $\kappa \in \mathbb{R}^d$. For this purpose we consider the function $V_1(\mathbf{x}) = (C\mathbf{x},\mathbf{x})$. Applying the generating operator L of the process $\hat{\mathbf{x}}^{\tau,\kappa}(t)$ to this function, we get that

$$\begin{aligned}
LV_1(\hat{\mathbf{x}}^{\tau,\kappa}) &= \mathrm{E}\Bigg[C\left(\mathbf{x} - \frac{a(\hat{\mathbf{f}}(\mathbf{x}) + \mathbf{q}(t,\bar{\mathbf{x}}(t))) + \hat{\mathbf{W}}(t+1,\bar{x}(t))}{t+1}\right), \\
&\quad \mathbf{x} - \frac{a(\hat{\mathbf{f}}(\mathbf{x}) + \mathbf{q}(t,\bar{\mathbf{x}}(t))) + \hat{\mathbf{W}}(t+1,\bar{x}(t))}{t+1}\Bigg] - (C\mathbf{x},\mathbf{x}) \\
&= -\frac{2a}{t+1}\mathrm{E}(C\mathbf{x},\hat{\mathbf{f}}(\mathbf{x}) + \mathbf{q}(t,\bar{\mathbf{x}}(t)) + \hat{\mathbf{W}}(t+1,\bar{\mathbf{x}})) \\
&\quad + \frac{a^2\mathrm{E}(C(\hat{\mathbf{f}}(\mathbf{x}) + \mathbf{q}(t,\bar{\mathbf{x}}(t))),\hat{\mathbf{f}}(\mathbf{x}) + \mathbf{q}(t,\bar{\mathbf{x}}(t)))}{(t+1)^2} + \frac{a^2\mathrm{E}\hat{\mathbf{W}}^2(t+1,\bar{\mathbf{x}})}{(t+1)^2} \\
&\leq -\frac{2a}{t+1}(C\hat{\mathbf{f}}(\mathbf{x}),\mathbf{x}) + \frac{2a}{t+1}\mathrm{E}|(C\mathbf{q}(t,\bar{\mathbf{x}}(t)),\mathbf{x})| \\
&\quad + \frac{a^2\|C\|\mathrm{E}(\|\hat{\mathbf{f}}(\mathbf{x}) + \mathbf{q}(t,\bar{\mathbf{x}}(t))\|^2 + 1)}{(t+1)^2}.
\end{aligned}$$

Recall that for some positive constants M and M' almost surely

$$\|\mathbf{q}(t,\bar{\mathbf{x}}(t))\| \leq c(t) = M\left(\lambda^t + \frac{a}{t+1}\sum_{s=1}^{t}\lambda^{t-s}\right) \leq M', \tag{4.14}$$

for $t \in \mathbb{Z}^+$. According to the definition of the function $\hat{\mathbf{f}}$ and Assumption 4.3.1 respectively, $\|\hat{\mathbf{f}}(\mathbf{x})\| = \beta\|\mathbf{x}\|$ for $\|\mathbf{x}\| \geq \varepsilon$ and $\hat{\mathbf{f}}(\mathbf{x})$ is bounded for $\|\mathbf{x}\| < \varepsilon$. Thus, taking into account (4.14), we conclude that there exists a positive constant k such that $\mathrm{E}(\|\hat{\mathbf{f}}(x) + \mathbf{q}(t, \bar{\mathbf{x}}(t))\|^2 + 1) \leq k(\|\mathbf{x}\|^2 + 1)$. Hence, using the Cauchy–Schwarz inequality and the fact that $\|\mathbf{x}\| \leq 1 + \|\mathbf{x}\|^2$, we obtain

$$\begin{aligned} LV_1(\hat{\mathbf{x}}^{\tau,\kappa}) &\leq -\frac{2a}{t+1}(C\hat{\mathbf{f}}(\mathbf{x}), \mathbf{x}) + \frac{2ac(t)\|C\|(1 + \|\mathbf{x}\|^2)}{t+1} + \frac{k_2(1 + \|\mathbf{x}\|^2)}{(t+1)^2} \\ &\leq -\frac{2a}{t+1}(C\hat{\mathbf{f}}(\mathbf{x}), \mathbf{x}) + \frac{k_1 c(t)(1 + (C\mathbf{x}, \mathbf{x}))}{t+1} + \frac{k_3(1 + (C\mathbf{x}, \mathbf{x}))}{(t+1)^2}, \end{aligned}$$

where k_1, k_2, and k_3 are some positive constants. Taking into account that $(C\hat{\mathbf{f}}(\mathbf{x}), \mathbf{x}) \geq \beta(C\mathbf{x}, \mathbf{x})$ and $2a\beta > 1$, we conclude that

$$LV_1(\hat{\mathbf{x}}^{\tau,\kappa}) \leq -\frac{p_1 V_1(\hat{\mathbf{x}}^{\tau,\kappa})}{t+1} + (1 + V_1(\hat{\mathbf{x}}^{\tau,\kappa}))\left(\frac{k_1 c(t)}{t+1} + \frac{k_3}{(t+1)^2}\right),$$

where $p_1 > 1$. According to (4.14), there exists such constant $p \in (1, p_1)$ that

$$-\frac{p_1}{t+1} + \frac{k_1 c(t)}{t+1} \geq -\frac{p}{t+1}, \quad -\frac{p_1}{t+1} + \frac{k_3}{(t+1)^2} \geq -\frac{p}{t+1}$$

are fulfilled simultaneously, if $t > T$, where $T \geq \tau$ is some finite sufficiently large constant. Thus, for some $p > 1$ and $T \geq 0$, we have

$$LV_1(\hat{\mathbf{x}}^{\tau,\kappa}) \leq -\frac{pV_1(\mathbf{x})}{t+1} + \frac{k_1 c(t)}{t+1} + \frac{k_3}{(t+1)^2}, \text{ for any } t > T.$$

Therefore, for $t > T$

$$\mathrm{E}V_1(\hat{\mathbf{x}}^{\tau,\kappa}(t+1)) - \mathrm{E}V_1(\hat{\mathbf{x}}^{\tau,\kappa}(t)) \leq -\frac{p\mathrm{E}V_1(\hat{\mathbf{x}}^{\tau,\kappa})}{t+1} + \frac{k_1 c(t)}{t+1} + \frac{k_3}{(t+1)^2}.$$

Using the above inequality, we have

$$\begin{aligned} \mathrm{E}V_1(\hat{\mathbf{x}}^{\tau,\kappa}(t+1)) \leq \mathrm{E}V_1(\hat{\mathbf{x}}^{\tau,\kappa}(T)) &\prod_{r=T}^{t}\left(1 - \frac{p}{r+1}\right) \\ &+ k_1 \sum_{r=T}^{t} \frac{1}{(r+1)^2} \prod_{m=r+1}^{t}\left(1 - \frac{p}{m+1}\right) \\ &+ k_3 \sum_{r=T}^{t} \frac{c(r)}{r+1} \prod_{m=r+1}^{t}\left(1 - \frac{p}{m+1}\right). \end{aligned}$$

Since

$$\prod_{m=r+1}^{t}\left(1-\frac{1}{m+1}\right)\leq \exp\left(-\sum_{m=r+1}^{t}\frac{1}{m+1}\right),$$

$$\sum_{m=r+1}^{t}\frac{1}{m+1}\geq \int_{r+1}^{t}\frac{1}{m+1}dm=\frac{\ln(t+1)}{\ln(r+1)},$$

we get that for some $k_4 > 0$

$$\prod_{m=r+1}^{t}\left(1-\frac{p}{m}\right)\leq k_4\left(\frac{r+1}{t+1}\right)^p,$$

and, hence, for some $k_5, k_6 > 0$

$$\sum_{r=T}^{t}\frac{1}{(r+1)^2}\prod_{m=r+1}^{t}\left(1-\frac{p}{m+1}\right)\leq\frac{k_5}{t+1},$$

$$\sum_{r=T}^{t}\frac{c(r)}{r+1}\prod_{m=r+1}^{t}\left(1-\frac{p}{m+1}\right)\leq\frac{k_6}{t+1}.$$

The last inequalities are due to (4.14) and the fact that $\sum_{r=1}^{t} r^{p-2} = O(t^{p-1})$.[7] Thus, we finally conclude that

$$\mathrm{E}V_1(\hat{\mathbf{x}}^{\tau,\kappa}(t+1)) = O\left(\frac{1}{t^p}\right)+O\left(\frac{1}{t}\right)=O\left(\frac{1}{t}\right),$$

that implies

$$\mathrm{E}\|\hat{\mathbf{x}}^{\tau,\kappa}(t)\|^2 = O\left(\frac{1}{t}\right)\ \text{as } t\to\infty \text{ for any } \tau\geq 0, \kappa\in\mathbb{R}^d, \tag{4.15}$$

since $V_1(\mathbf{x}) = (C\mathbf{x},\mathbf{x})$ and the matrix C is positive definite.

Since we assumed that $\Pr\{\lim_{s\to\infty}\bar{\mathbf{x}}(s) = \mathbf{0}\} > 0$, then the rate of convergence of $\mathrm{E}\{\|\bar{\mathbf{x}}(t)\|^2 \mid \lim_{s\to\infty}\bar{\mathbf{x}}(s) = \mathbf{0}\}$ and $\mathrm{E}\{\|\bar{\mathbf{x}}(t)\|^2\mathbb{1}(\lim_{s\to\infty}\bar{\mathbf{x}}(s) = \mathbf{0})\}$ will be the same, where $\mathbb{1}\{\cdot\}$ denotes the event indicator function. But

$$\mathrm{E}(\|\bar{\mathbf{x}}(t)\|^2\mathbb{1}\{\lim_{s\to\infty}\bar{\mathbf{x}}(s)=\mathbf{0}\}) = \int_{\mathbb{R}} x^2 d\left(\Pr\{\|\bar{\mathbf{x}}(t)\|^2\leq x, \lim_{s\to\infty}\bar{\mathbf{x}}(s)=\mathbf{0}\}\right).$$

[7] This estimation can be obtained by considering the sum $\sum_{r=1}^{t} r^{p-2}$ the low sum of the corresponding integrals for two cases: $p\geq 2$, $1<p<2$.

Thus, to get asymptotic behavior of $E\{\|\bar{\mathbf{x}}(t)\|^2 \mid \lim_{s\to\infty}\bar{\mathbf{x}}(s) = \mathbf{0}\}$ we need to analyze the asymptotics of the distribution function $\Pr\{\|\bar{\mathbf{x}}(t)\|^2 \le x, \lim_{s\to\infty}\bar{\mathbf{x}}(s) = \mathbf{0}\}$. For this purpose we introduce the following events:

$$\Theta_{u,\tilde{\varepsilon}} = \{\|\bar{\mathbf{x}}(u)\| < \tilde{\varepsilon}\},\ \Omega_{u,\tilde{\varepsilon}} = \{\|\bar{\mathbf{x}}(m)\| < \tilde{\varepsilon},\ m \ge u\}.$$

Since $\bar{\mathbf{x}}(t)$ converges to $\mathbf{x}^* = \mathbf{0}$ with some positive probability, then for any $\sigma > 0$ there exist $\tilde{\varepsilon}(\sigma)$ and $u(\sigma)$ such that for any $u \ge u(\sigma)$ the following is hold:

$$\Pr\{\{\lim_{s\to\infty}\bar{\mathbf{x}}(s) = \mathbf{0}\}\ \Delta\ \Omega_{u,\tilde{\varepsilon}(\sigma)}\} < \sigma,$$

$$\Pr\{\{\lim_{s\to\infty}\bar{\mathbf{x}}(s) = \mathbf{0}\}\ \Delta\ \Theta_{u,\tilde{\varepsilon}(\sigma)}\} < \sigma,$$

$$\Pr\{\Theta_{u,\tilde{\varepsilon}(\sigma)}\ \Delta\ \Omega_{u,\tilde{\varepsilon}(\sigma)}\} < \sigma,$$

where Δ denotes the symmetric difference. Hence, choosing $\varepsilon' = \min(\tilde{\varepsilon}, \varepsilon)$, we get that for $u \ge u(\sigma)$

$$\begin{aligned}
\Pr\{\|\bar{\mathbf{x}}(t)\|^2 \le x,\ \lim_{s\to\infty}\bar{\mathbf{x}}(s) = \mathbf{0}\} &\le \Pr\{\|\bar{\mathbf{x}}(t)\|^2 \le x, \{\lim_{s\to\infty}\bar{\mathbf{x}}(s) = \mathbf{0}\} \cup \{\Omega_{u,\varepsilon'}\}\} \\
&\le \Pr\{\|\bar{\mathbf{x}}(t)\|^2 \le x, \Omega_{u,\varepsilon'}\} + \sigma = \Pr\{\|\hat{\mathbf{x}}^{u,\bar{\mathbf{x}}(u)}(t)\|^2 \le x, \Omega_{u,\varepsilon'}\} + \sigma \\
&\le \Pr\{\|\hat{\mathbf{x}}^{u,\bar{\mathbf{x}}(u)}(t)\|^2 \le x, \Theta_{u,\varepsilon'}\} + 2\sigma.
\end{aligned}$$

Taking into account the Markovian property of the process $\{\hat{\mathbf{x}}^{u,\bar{\mathbf{x}}(u)}(t)\}$, we conclude from the inequality above that

$$\begin{aligned}
\overline{\lim_{t\to\infty}} \Pr\{\|\bar{\mathbf{x}}(t)\|^2 \le x,\ \lim_{s\to\infty}\bar{\mathbf{x}}(s) = \mathbf{0}\} &\le \overline{\lim_{t\to\infty}} \Pr\{\|\hat{\mathbf{x}}^{u,\bar{\mathbf{x}}(u)}(t)\|^2 \le x, \Theta_{u,\varepsilon'}\} + 2\sigma \\
&= \overline{\lim_{t\to\infty}} \Pr\{\|\hat{\mathbf{x}}^{u,\bar{\mathbf{x}}(u)}(t)\|^2 \le x\} \Pr\{\Theta_{u,\varepsilon'}\} + 2\sigma \\
&\le \overline{\lim_{t\to\infty}} \Pr\{\|\hat{\mathbf{x}}^{u,\bar{\mathbf{x}}(u)}(t)\|^2 \le x\} \Pr\{\lim_{s\to\infty}\bar{\mathbf{x}}(s) = \mathbf{0}\} + 3\sigma. \qquad (4.16)
\end{aligned}$$

Analogously,

$$\begin{aligned}
&\underline{\lim_{t\to\infty}} \Pr\{\|\bar{\mathbf{x}}(t)\|^2 \le x,\ \lim_{s\to\infty}\bar{\mathbf{x}}(s) = \mathbf{0}\} \\
&\quad \le \underline{\lim_{t\to\infty}} \Pr\{\|\hat{\mathbf{x}}^{u,\bar{\mathbf{x}}(u)}(t)\|^2 \le x\} \Pr\{\lim_{s\to\infty}\bar{\mathbf{x}}(s) = \mathbf{0}\} + 3\sigma. \qquad (4.17)
\end{aligned}$$

Similarly, we can obtain that

$$\begin{aligned}
&\Pr\{\|\bar{\mathbf{x}}(t)\|^2 \leq x, \lim_{s\to\infty} \bar{\mathbf{x}}(s) = \mathbf{0}\} \\
&\geq \Pr\{\|\bar{\mathbf{x}}(t)\|^2 \leq x, \Omega_{u,\varepsilon'} \setminus \{\Omega_{u,\varepsilon'} \cap \{\lim_{s\to\infty} \bar{\mathbf{x}}(s) = \mathbf{0}\}\}\} \\
&\geq \Pr\{\|\hat{\mathbf{x}}^{u,\bar{\mathbf{x}}(u)}(t)\|^2 \leq x, \Theta_{u,\varepsilon'}\} - 2\sigma,
\end{aligned}$$

and, hence,

$$\begin{aligned}
&\overline{\lim_{t\to\infty}} \Pr\{\|\bar{\mathbf{x}}(t)\|^2 \leq x, \lim_{s\to\infty} \bar{\mathbf{x}}(s) = \mathbf{0}\} \\
&\geq \overline{\lim_{t\to\infty}} \Pr\{\|\hat{\mathbf{x}}^{u,\bar{\mathbf{x}}(u)}(t)\|^2 \leq x\} \Pr\{\lim_{s\to\infty} \bar{\mathbf{x}}(s) = \mathbf{0}\} - 3\sigma, \qquad (4.18)
\end{aligned}$$

$$\begin{aligned}
&\underline{\lim}_{t\to\infty} \Pr\{\|\bar{\mathbf{x}}(t)\|^2 \leq x, \lim_{s\to\infty} \bar{\mathbf{x}}(s) = \mathbf{0}\} \\
&\geq \underline{\lim}_{t\to\infty} \Pr\{\|\hat{\mathbf{x}}^{u,\bar{\mathbf{x}}(u)}(t)\|^2 \leq x\} \Pr\{\lim_{s\to\infty} \bar{\mathbf{x}}(s) = \mathbf{0}\} - 3\sigma. \qquad (4.19)
\end{aligned}$$

Since σ can be chosen arbitrary small, (4.16)–(4.19) imply that

$$\overline{\lim_{t\to\infty}} \Pr\{\|\bar{\mathbf{x}}(t)\|^2 \leq x, \lim_{s\to\infty} \bar{\mathbf{x}}(s) = \mathbf{0}\} = \overline{\lim_{t\to\infty}} \Pr\{\|\hat{\mathbf{x}}^{u,\bar{\mathbf{x}}(u)}(t)\|^2 \leq x\} \Pr\{\lim_{s\to\infty} \bar{\mathbf{x}}(s) = \mathbf{0}\},$$

$$\underline{\lim}_{t\to\infty} \Pr\{\|\bar{\mathbf{x}}(t)\|^2 \leq x, \lim_{s\to\infty} \bar{\mathbf{x}}(s) = \mathbf{0}\} = \underline{\lim}_{t\to\infty} \Pr\{\|\hat{\mathbf{x}}^{u,\bar{\mathbf{x}}(u)}(t)\|^2 \leq x\} \Pr\{\lim_{s\to\infty} \bar{\mathbf{x}}(s) = \mathbf{0}\}.$$

Thus, we conclude that asymptotic behavior of $\Pr\{\|\bar{\mathbf{x}}(t)\|^2 \leq x, \lim_{s\to\infty} \bar{\mathbf{x}}(s) = \mathbf{0}\}$ coincides with the asymptotics of $\Pr\{\|\hat{\mathbf{x}}^{u,\bar{\mathbf{x}}(u)}(t)\|^2 \leq x\} \Pr\{\lim_{s\to\infty} \bar{\mathbf{x}}(s) = \mathbf{0}\}$. Now we can use (4.15) and the fact that $\Pr\{\lim_{s\to\infty} \bar{\mathbf{x}}(s) = \mathbf{0}\} \in (0, 1]$ to get

$$E\{\|\bar{\mathbf{x}}(t)\|^2 \mid \lim_{s\to\infty} \bar{\mathbf{x}}(s) = \mathbf{0}\} = O\left(\frac{1}{t}\right) \text{ as } t \to \infty.$$

□

4.3.5 *Simulation Results: Illustrative Example and Congestion Routing Problem*

1. Let us consider minimization of the following scalar function:

$$F(z) = F_1(z) + F_2(z) + F_3(z),$$

over a network of three agents. We assume that the function F_i is known only to the agent i, $i = 1, 2, 3$, and

$$F_1(z) = \begin{cases} (z^3 - 16x)(z+2), \text{ if } |z| \le 10, \\ 4248z - 32400, \text{ if } z > 10, \\ -3112z - 25040, \text{ if } z < -10. \end{cases}$$

$$F_2(z) = \begin{cases} (0.5z^3 + z^2)(z-4), \text{ if } |z| \le 10, \\ 1620z - 12600, \text{ if } z > 10, \\ -2220z - 16600, \text{ if } z < -10. \end{cases}$$

$$F_3(z) = \begin{cases} (z+2)^2(z-4), \text{ if } |z| \le 10, \\ 288z - 2016, \text{ if } z > 10, \\ 288z - 1984, \text{ if } z < -10. \end{cases}$$

Thus, the plot of the function F on the interval $z \in [-6, 6]$ is represented by Fig. 4.1. Outside this interval the function F has no local minima.[8]

Let us assume that the communication graph $G(t)$ has a switching topology and at the even time steps, $t = 0, 2, 4, \ldots$, it is $G(t) = ([3], \{1 \to 2, 2 \to 3\})$ and at the odd ones, $t = 1, 3, 5, \ldots$, it is $G(t) = ([3], \{2 \to 1, 3 \to 2\})$. Thus, the graph $G(t)$ is 2-strongly connected. Moreover, Assumptions 4.3.1–4.3.3 and 4.3.5 hold for the functions $\{F_i(z)\}_i$ and $F(z)$. That is why we can apply the perturbed push-sum algorithm to guarantee almost sure convergence to a local minimum, namely either to $z = -2.49$ or to $z = 2.62$. The performance of the algorithm for the following initial values of the auxiliary variables $x_1(0) = 0$, $x_2(0) = 0$, $x_3(0) = 0$ and $x_1(0) = -1$, $x_2(0) = -1.2$, $x_3(0) = -1.1$ are demonstrated by Figs. 4.2, 4.3, 4.4, 4.5, 4.6, and 4.7, respectively. We can see that in the first case the algorithm converges to the global minimum $z = 2.62$. In the second case, although the initial estimations are close to a critical point (local maximum of F), namely $z = -1.12$, the algorithm converges to the local minimum $z = -2.49$.

2. Another problem considered here is a special case of the congestion routing problem from Example 2.3.1. We consider again a network of three agents with the same communication topology $G(t)$ as above. Let us assume that each agent can transfer not more than 1000 units of demand from a start point to a destination. However, agents have preferences to send definite amount of demand. For example, the agent 1 gets the highest profit, if she sends 66 units of

[8]Note that the problem $F(z) \to \min_{\mathbb{R}}$ is equivalent to the problem of minimization of the function $F_0(z) = (z^3 - 16z)(z+2) + (0.5z^3 + z^2)(z-4) + (z+2)^2(z-4)$ on the interval $[-10, 10]$, where $F_0(z)$ is extended on the whole $\mathbb{R}$ to satisfy Assumption 4.3.1.

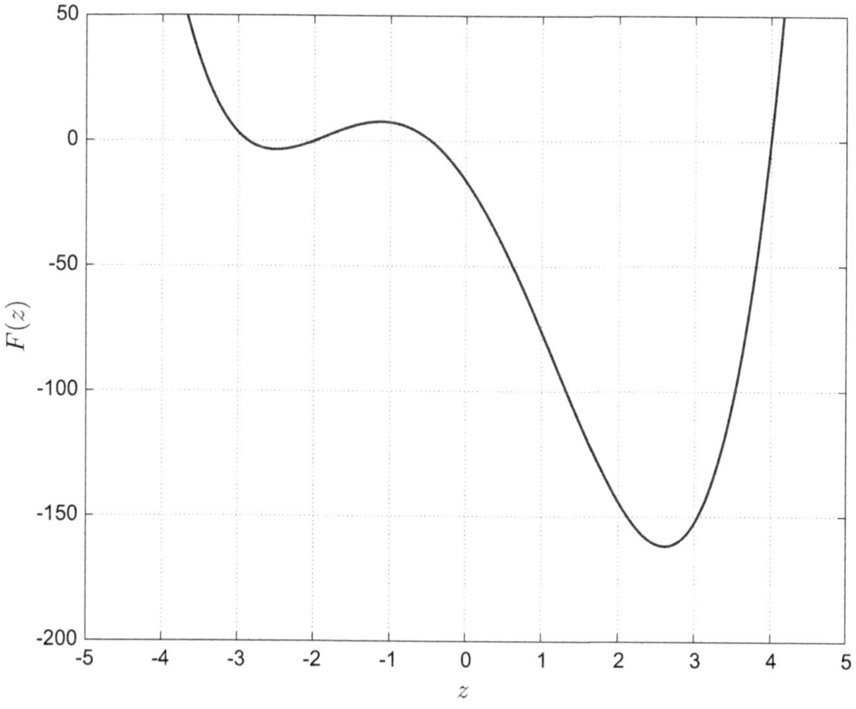

Fig. 4.1 Global objective function $F(z)$

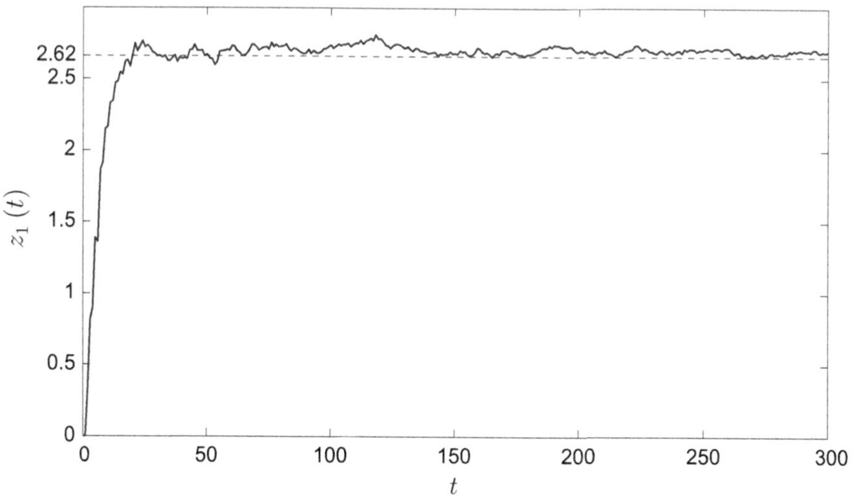

Fig. 4.2 The value $z_1(t)$ in the run of the perturbed push-sum algorithm (given $x_1(0) = 0, x_2(0) = 0, x_3(0) = 0$)

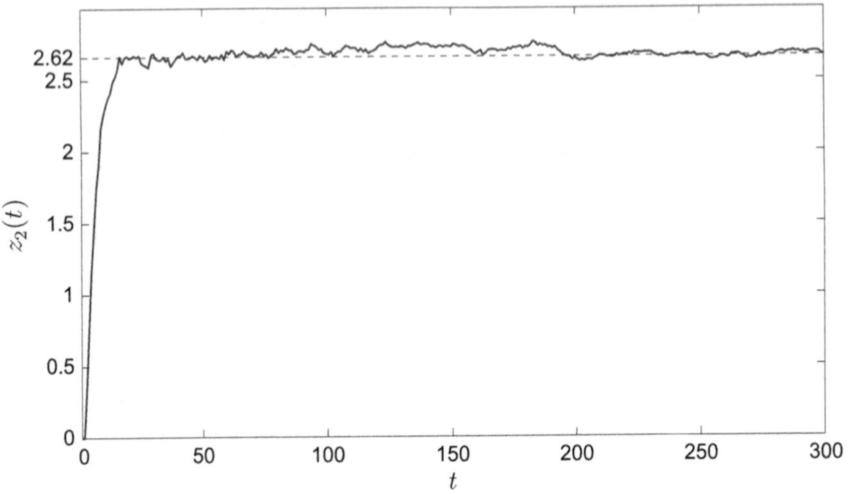

Fig. 4.3 The value $z_2(t)$ in the run of the perturbed push-sum algorithm (given $x_1(0) = 0, x_2(0) = 0, x_3(0) = 0$)

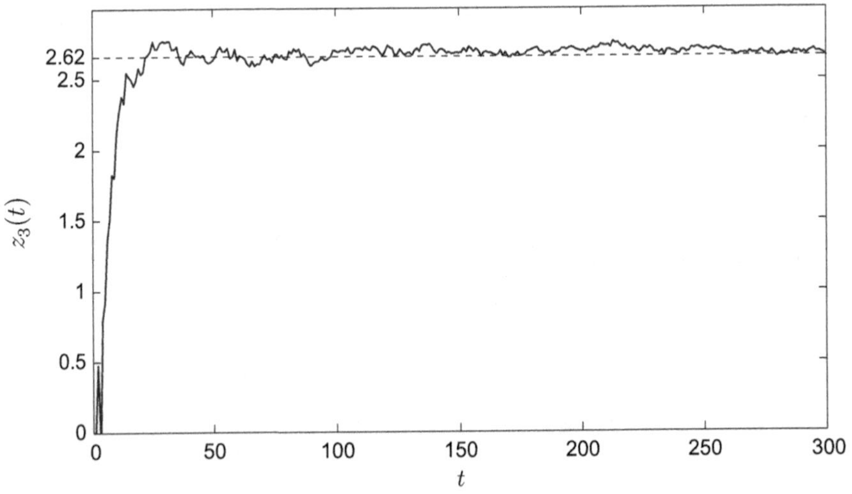

Fig. 4.4 The value $z_3(t)$ in the run of the perturbed push-sum algorithm (given $x_1(0) = 0, x_2(0) = 0, x_3(0) = 0$)

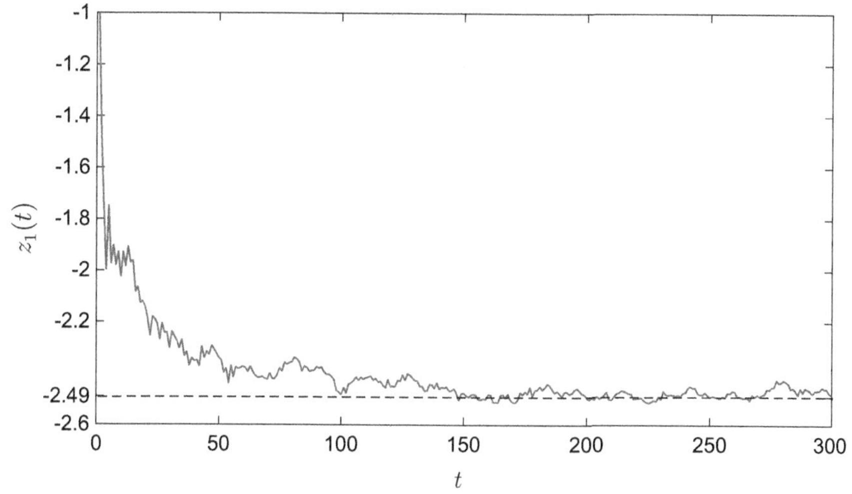

Fig. 4.5 The value $z_1(t)$ in the run of the perturbed push-sum algorithm (given $x_1(0) = -1$, $x_2(0) = -1.2, x_3(0) = -1.1$)

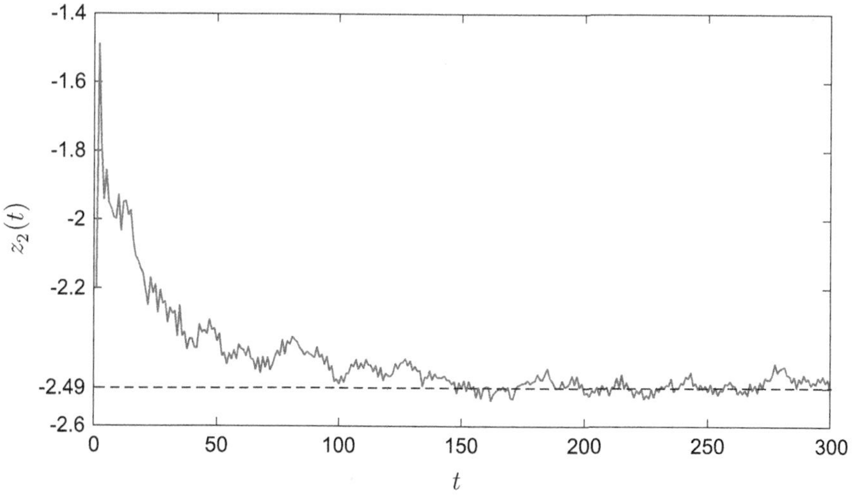

Fig. 4.6 The value $z_2(t)$ in the run of the perturbed push-sum algorithm (given $x_1(0) = -1$, $x_2(0) = -1.2, x_3(0) = -1.1$)

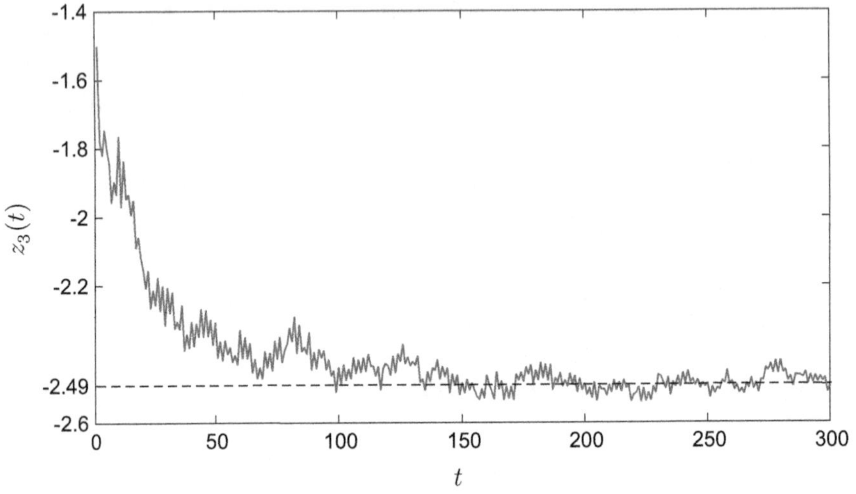

Fig. 4.7 The value $z_3(t)$ in the run of the perturbed push-sum algorithm (given $x_1(0) = -1$, $x_2(0) = -1.2$, $x_3(0) = -1.1$)

her demand and her utility function, given she transfers z'_1 units, is

$$u_1(z'_1) = -\frac{(z'_1 - 66)^2}{z'_1}.$$

Similarly, other two agents have their preferences and their utility functions are

$$u_2(z'_2) = -\frac{(z'_2 - 70)^2}{z'_2}$$

and

$$u_3(z'_3) = -\frac{(z'_3 - 38)^2}{z'_3}$$

respectively. Note that the agents get a highly negative profit if they transfer some small amount of data ($z'_i \sim 0$). The costs associated with the congestion in the system are different for the agents, namely

$$p_1(\mathbf{z}') = 0.001z'_1((z'_1 + z'_2 + z'_3)^2 + 5(z'_1 + z'_2 + z'_3) + 5),$$
$$p_2(\mathbf{z}') = 0.001z'_2((z'_1 + z'_2 + z'_3)^2 + 1),$$
$$p_3(\mathbf{z}') = 0.001z'_3((z'_1 + z'_2 + z'_3)^2 + 5),$$

where $\mathbf{z}' = (z_1', z_2', z_3')$. Thus, the optimization problem in the system under consideration can be formulated as follows:

$$C(\mathbf{z}') = \sum_{i=1}^{3} c_i(\mathbf{z}') = \sum_{i=1}^{3} (p_i(\mathbf{z}') - u_i(z_i')) \to \min, \tag{4.20}$$

$$\text{s.t. } z_1' \in [0, 100], z_2' \in [0, 100], z_3' \in [0, 100]. \tag{4.21}$$

To be able to deal with the unconstrained problem, we substitute in the constrained problem above z_i' by e^{z_i}, $z_i \in (-\infty, \ln(1000)]$. Note that the functions $c_i(\mathbf{z})$, $i = 1, 2, 3$, on the set $R_{1,\text{out}} = \cup_{i=1}^{3}\{z_i > \ln(1000)\}$ and $R_{2,\text{out}} = \cup_{i=1}^{3}\{z_i < -1000\}$ can be redefined by such functions $\tilde{c}_i$, $i = 1, 2, 3$, that there are no local minima of the resulting function $\tilde{C}(\mathbf{z}) = \sum_{i=1}^{3} \tilde{c}_i(\mathbf{z})$ on $R_{1,\text{out}} \cup R_{2,\text{out}}$ and Assumptions 4.3.1–4.3.3 hold for the functions $\{\tilde{c}_i(\mathbf{z})\}_i$ and $\tilde{C}(\mathbf{z})$. The run of the push-sum algorithm applied to the problem $\tilde{C}(\mathbf{z}) \to \min_{\mathbb{R}}$ is demonstrated by Figs. 4.8 and 4.9, given the following initial values for the auxiliary variables: $\mathbf{x}_1(0) = (4, 1, 2)$, $\mathbf{x}_2(0) = (1, 1, 2)$, $\mathbf{x}_3(0) = (1, 3, 1)$ and $\mathbf{x}_1(0) = (0, -1, -2)$, $\mathbf{x}_2(0) = (10, 0, 1)$, $\mathbf{x}_3(0) = (0, 1, -8)$, respectively. The figures show the change of the values of the function $\tilde{C}(\mathbf{z})$ during the run of the algorithm. We can see that in both cases the initial decay of the global cost is very fast and it takes only a small number of iterations for the agents to get close to the cost's optimal value 425.36. We remind here that the optimization problem $\tilde{C}(\mathbf{z}) \to \min_{\mathbb{R}}$ is equivalent to the original one (4.20).

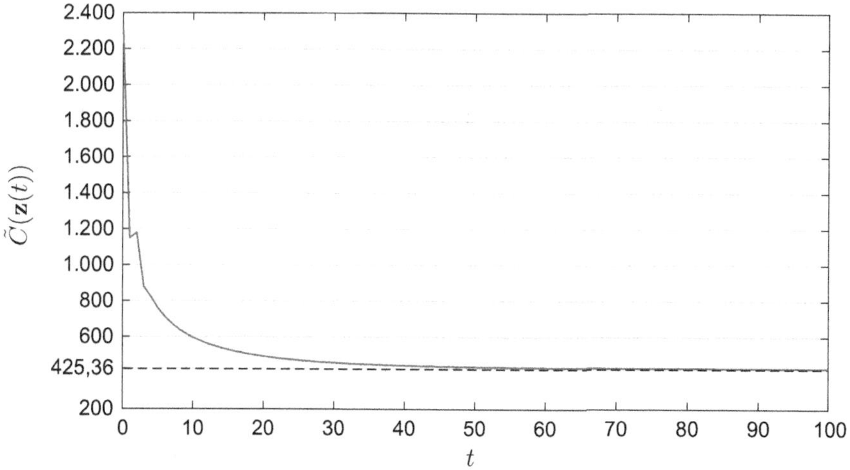

Fig. 4.8 The value $\tilde{C}$ in the run of the perturbed push-sum algorithm (given $\mathbf{x}_1(0) = (4, 1, 2)$, $\mathbf{x}_2(0) = (1, 1, 2)$, $\mathbf{x}_3(0) = (1, 3, 1)$)

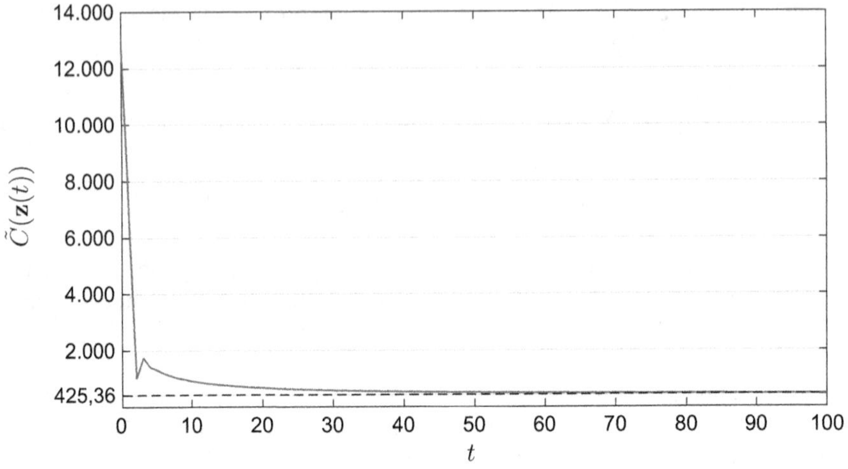

Fig. 4.9 The value $\widetilde{C}$ in the run of the perturbed push-sum algorithm (given $\mathbf{x}_1(0) = (0, -1, -2)$, $\mathbf{x}_2(0) = (10, 0, 1)$, $\mathbf{x}_3(0) = (0, 1, -8)$)

4.4 Communication-Based Memoryless Learning in Potential Games

This section adapts the push-sum algorithm, which was discussed previously in context of distributed optimization, to learning in potential games. According to the rule of the algorithm (see, for example, (4.3)), nodes update the arguments $\mathbf{z}_i$ based on the information currently obtained from their neighbors. Thus, this algorithm can be implemented in a multi-agent system, where agents have no memory.

The focus is on a multi-agent system with N agents, taking actions from $\mathbb{R}$, and the objective function $\phi : \mathbb{R}^N \to \mathbb{R}$. Thus, the problem to be solved in such multi-agent system can be formulated as follows:

$$\max \phi(\boldsymbol{a}),$$
$$\text{s.t. } \boldsymbol{a} \in \mathbb{R}^N. \tag{4.22}$$

The goal of this subsection is to find a local solution of the problem above.

According to the discussion in Sect. 2.2, this system is modeled by means of a continuous action potential game $(N, \{A_i\}, \{U_i\}, \phi)$ with $A_i = \mathbb{R}$, $i \in [N]$. Before describing the information available in the system and formulating the learning algorithm, we introduce the following assumptions on the potential function, under which the learning algorithm will be analyzed.

Assumption 4.4.1 *The function ϕ in the problem* (4.22) *has finitely many critical points which lie on some compact subset of $\mathbf{A} = \mathbb{R}^N$.*

Let $\boldsymbol{A}_0$ and $\boldsymbol{A}^*$ denote the set of critical points and local maxima of the function ϕ on $\boldsymbol{A}$, respectively ($\boldsymbol{A}^* \subseteq \boldsymbol{A}_0$).

Assumption 4.4.2 *The gradient* $\nabla\phi$ *of the function* ϕ *exists and is bounded on* $\boldsymbol{A} = \mathbb{R}^N$.

Assumption 4.4.3 *The function* $-\phi$ *is coercive, i.e.,* $\phi(\boldsymbol{a}) \to -\infty$ *as* $\|\boldsymbol{a}\| \to \infty$.

Assumption 4.4.4 *For any point* $\boldsymbol{a}' \in \boldsymbol{A}_0 \backslash \boldsymbol{A}^*$ *that is not a local maximum of* ϕ *there exists a symmetric positive definite matrix* $C(\boldsymbol{a}')$ *such that* $(\nabla\phi(\boldsymbol{a}), C(\boldsymbol{a}')(\boldsymbol{a} - \boldsymbol{a}')) \geq 0$ *for any* $\boldsymbol{a} \in U(\boldsymbol{a}')$, *where* $U(\boldsymbol{a}')$ *is some open neighborhood of* $\boldsymbol{a}'$.

Let us consider a system, where agents can communicate and exchange information only with their neighbors. Additionally to access to such communication, each agent i can estimate the partial derivative $\frac{\partial U_i}{\partial a^i}$ at any given point. The communication topology is defined by a directed graph $G(t)$. We deal with the time-varying topology $G(t)$ satisfying the following assumption:

Assumption 4.4.5 *The sequence of communication graphs* $\{G(t)\}$ *is S-strongly connected (see Definition 4.3.1).*

As before $N_i^{\text{in}}(t)$ and $N_i^{\text{out}}(t)$ is used to denote the in- and out-neighborhoods of the agent i at time t. Each agent i is represented by the corresponding node in $G(t)$ and is always considered to be an in- and out-neighbor of itself. We use $d_i(t)$ to denote the out-degree of the agent i, and we assume that every agent knows its out-degree at every time t.

Thus, the learning algorithm to be applied to the game should be of the following type:

$$a^i(t+1) = F_i\left(N_i^{\text{out}}(t), \frac{\partial U_i}{\partial a^i}\right), \tag{4.23}$$

where $a^i(t+1)$ is the action the player i chooses at the moment t and plays at the moment $t+1$.

Another assumption, under which the learning algorithm will be analyzed, concerns the continuous property of the gradient $\nabla\phi$:

Assumption 4.4.6 *The gradient* $\nabla\phi$ *is coordinate-wise Lipschitz continuous on* $\boldsymbol{A} = \mathbb{R}^N$, *i.e., there exists such positive constant L that* $\left|\frac{\partial\phi(\boldsymbol{a}_1)}{\partial a^i} - \frac{\partial\phi(\boldsymbol{a}_2)}{\partial a^i}\right| \leq L\|\boldsymbol{a}_1 - \boldsymbol{a}_2\|$ *for any* $i \in [N]$ *and* $\boldsymbol{a}_1, \boldsymbol{a}_2 \in \mathbb{R}^N$.

We remind that the goal of the agents is to find a local solution of the problem (4.22) modeled as a potential game $(N, \{A_i = \mathbb{R}\}, \{U_i\}, \phi)$. To reach this goal agents adapt the *push-sum protocol*, which was discussed in the previous section, as follows. At every moment of time $t \in \mathbb{Z}^+$ each agent i maintains, besides her current action a^i, vector variables $\mathbf{a}_i(t) = (\mathrm{a}_i^1, \dots, \mathrm{a}_i^N), \boldsymbol{x}_i(t), \boldsymbol{w}_i(t) \in \mathbb{R}^N$, as well as a scalar variable $y_i(t) : y_i(0) = 1$. These quantities are updated according to the *communication-based algorithm* as follows:

$$\begin{aligned}
\boldsymbol{w}_i(t+1) &= \sum_{j \in N_i^{\text{in}}(t)} \frac{\boldsymbol{x}_j(t)}{d_j(t)}, \\
y_i(t+1) &= \sum_{j \in N_i^{\text{in}}(t)} \frac{y_j(t)}{d_j(t)}, \\
\mathbf{a}_i(t+1) &= \frac{\boldsymbol{w}_i(t+1)}{y_i(t+1)}, \\
\boldsymbol{x}_i(t+1) &= \boldsymbol{w}_i(t) + \gamma(t+1)[\mathbf{f}_i(\mathbf{a}_i(t+1)) + \boldsymbol{\xi}_i(t)], \\
a^i(t+1) &= \mathrm{a}_i^i(t+1),
\end{aligned} \tag{4.24}$$

where $\{\gamma(t)\}_t$ is a specified sequence of time steps,

$$\mathbf{f}_i(\mathbf{a}_i(t+1)) = \left(0, \ldots, \frac{\partial U_i}{\partial a^i}(\mathbf{a}_i(t+1)), \ldots, 0\right)^T$$

is the N-dimensional vector with the unique non-zero coordinate, and $\{\boldsymbol{\xi}_i(t)\}_t$ is the sequence of independently and identically distributed (i.i.d.) random vectors taking values in $\mathbb{R}^N$ and satisfying the following assumption:

Assumption 4.4.7 *The coordinates $\xi_i^k(t)$, $\xi_i^l(t)$ are independent, $\|\xi_i^k(t)\|$ is bounded almost surely, and $\mathrm{E}(\xi_i^k(t)) = 0$ for all $t \in \mathbb{Z}^+$ and $i, k, l \in [N]$.*

The process above is a special case of the *general push-sum algorithm* (4.3) (Sect. 4.3.3), where $\mathbf{e}_i(t) = \mathbf{f}_i(\mathbf{a}_i(t+1)) + \boldsymbol{\xi}_i(t)$. Thus, the algorithm above can be considered the adaptation of (4.3) to learning in potential games.

In the following it will be demonstrated that, given Assumptions (4.4.1)–(4.4.7) and an appropriate choice of $\gamma(t)$, the algorithm (4.24) converges almost surely to a local maximum $\boldsymbol{a}^* \in \boldsymbol{A}^*$ of the potential function, namely $\Pr\{\lim_{t\to\infty} \mathbf{a}(t) = \boldsymbol{a}^*\} = 1$. Thus, taking into account the last equality in (4.24), we can conclude that the joint action $\boldsymbol{a}(t)$ converges with time to a local maximum of the potential function almost surely.

Firstly, let us notice that under the rules of the process (4.24) the average of the vector variables $\boldsymbol{x}_i(t), i \in [N]$, namely $\bar{\boldsymbol{x}}(t) = \frac{1}{N}\sum_{i=1}^N \boldsymbol{x}_i(t)$, is updated as follows:

$$\bar{\boldsymbol{x}}(t+1) = \bar{\boldsymbol{x}}(t) + \gamma(t+1)\left[\frac{\nabla\phi(\bar{\boldsymbol{x}}(t))}{N} + \boldsymbol{R}(t, \bar{\boldsymbol{x}}(t))\right] + \gamma(t+1)\boldsymbol{\xi}(t), \tag{4.25}$$

with

$$\boldsymbol{R}(t, \bar{\boldsymbol{x}}(t)) = \frac{1}{N}[\boldsymbol{F}(\mathbf{a}(t+1)) - \nabla\phi(\bar{\boldsymbol{x}}(t))],$$

where

$$F(\mathbf{a}(t+1)) = \sum_{i=1}^{N} \mathbf{f}_i(\mathbf{a}_i(t+1)) = \left(\frac{\partial U_1(\mathbf{a}_1(t+1))}{\partial a^1}, \ldots, \frac{\partial U_N(\mathbf{a}_N(t+1))}{\partial a^N}\right)$$

and $\boldsymbol{\xi}(t) = \frac{1}{N}\sum_{i=1}^{N} \boldsymbol{\xi}_i(t)$. Further the convergence properties of the procedure (4.25) will be analyzed to let us describe behavior of the communication-based algorithm (4.24). The main result of this subsection is formulated in the following theorem.

Theorem 4.4.1 *Let* $\Gamma = (N, \{A_i = \mathbb{R}\}, \{U_i\}, \phi)$ *be a* continuous *action potential game with* $A_i = \mathbb{R}$. *Let the parameter* $\gamma(t)$ *be such that* $\gamma(t) = O\left(\frac{1}{t^\nu}\right)$, *where* $\frac{1}{2} < \nu \leq 1$. *Then under Assumptions 4.4.1–4.4.7 the sequence* $\{\bar{\boldsymbol{x}}(t)\}$ *defined in* (4.25) *converges to a local maximum of the potential function* $\boldsymbol{a}^* \in \boldsymbol{A}^*$ *almost surely. Moreover, the joint action* $\boldsymbol{a}(t)$ *defined by* (4.24) *converges to* $\boldsymbol{a}^*$ *almost surely.*

Under Assumption 4.4.6 the gradient function $\nabla\phi(\boldsymbol{x})$ is a coordinate-wise Lipschitz function. This allows us to formulate the following lemma which will be needed for proof of Theorem 4.4.1.

Lemma 4.4.1 *Let* $\{\gamma(t)\}$ *be a non-increasing sequence such that* $\sum_{t=0}^{\infty}\gamma^2(t) < \infty$. *Then, under Assumptions 4.4.2, 4.4.5, and 4.4.7, there exists* $c(t)$ *such that the following holds for the process* (4.25):

$$\|\boldsymbol{R}(t, \bar{\boldsymbol{x}}(t))\| = \frac{1}{N}\|\boldsymbol{F}(\mathbf{a}(t+1)) - \nabla\phi(\bar{\boldsymbol{x}}(t))\| \leq c(t),$$

and $\sum_{t=0}^{\infty}\gamma(t+1)c(t) < \infty$.

The proofs of the lemma above and Theorem 4.4.1 follow the same line as in the proofs of Lemma 4.3.1 and Theorems 4.3.3–4.3.4, respectively. However, for the sake of consistency, also the proofs of Lemma 4.4.1 and Theorem 4.4.1 are presented here.

Proof of Lemma 4.4.1 Since the functions $\nabla\phi$ is a coordinate-wise Lipschitz one (Assumption 4.4.6) and taking into account $\frac{\partial\phi(\boldsymbol{a})}{\partial a^i} = \frac{\partial U_i(\boldsymbol{a})}{\partial a^i}$, we conclude that for any i

$$\begin{aligned}
\|\boldsymbol{F}(\mathbf{a}(t+1)) - \nabla\phi(\bar{\boldsymbol{x}}(t))\| &\leq \sum_{i=1}^{N}\left|\frac{\partial U_i(\mathbf{a}_i(t+1))}{\partial a^i} - \frac{\partial\phi(\boldsymbol{x}(t))}{\partial a^i}\right| \\
&= \sum_{i=1}^{N}\left|\frac{\partial\phi(\mathbf{a}_i(t+1))}{\partial a^i} - \frac{\partial\phi(\boldsymbol{x}(t))}{\partial a^i}\right| \\
&\leq L\sum_{i=1}^{N}\|\mathbf{a}_i(t+1) - \bar{\boldsymbol{x}}(t)\|.
\end{aligned}$$

Let $c(t) = \frac{L}{N}\sum_{i=1}^{N} \|\mathbf{a}_i(t+1) - \bar{\boldsymbol{x}}(t)\|$. Then, $\|\boldsymbol{R}(t, \bar{\boldsymbol{x}}(t))\| \le c(t)$. Taking into account Assumptions 4.4.2 and 4.4.7, we get

$$\sum_{t=1}^{\infty} \gamma(t)\|\mathbf{e}_i(t)\|_1 = \sum_{t=0}^{\infty} \gamma^2(t)\|\mathbf{f}_i(\mathbf{a}_i(t+1)) + \boldsymbol{\xi}_i(t)\|_1 \le k\sum_{t=0}^{\infty} \gamma^2(t) < \infty,$$

where k is some positive constant. Hence, according to Theorem 4.3.1,

$$\sum_{t=0}^{\infty} \gamma(t+1)c(t) = \frac{L}{N}\sum_{i=1}^{N}\sum_{t=0}^{\infty} \|\mathbf{a}_i(t+1) - \bar{\boldsymbol{x}}(t)\|\gamma(t+1) < \infty.$$

□

Proof of Theorem 4.4.1 To prove this statement we will use the general result formulated in Theorems 4.2.2 and 4.2.3. Let us notice that the process (4.25) under consideration represents a special case of the recursive procedure (4.1), where $\mathbf{X}(t) = \bar{\boldsymbol{x}}(t)$, $\alpha(t) = \beta(t) = \gamma(t)$, $\mathbf{f}(\mathbf{X}(t)) = \frac{\nabla\phi(\bar{\boldsymbol{x}}(t))}{N}$, $\mathbf{q}(t, \mathbf{X}(t)) = \boldsymbol{R}(t, \bar{\boldsymbol{x}}(t))$, and $\mathbf{W}(t, \mathbf{X}(t), \omega) = \boldsymbol{\xi}(t)$. In the following any k_j denotes some positive constant. Note that, according to the choice of $\gamma(t)$, Assumption 4.4.2, and Lemma 4.4.1, conditions (4.2.2) and (4.2.2) of Theorem 4.2.2 hold. Thus, it suffices to show that there exists a sample function $V(t, \boldsymbol{x})$ of the process (4.25) satisfying conditions (4.2.2) and (4.2.2). A natural choice for such a function is the (time-invariant) function $V(t, \boldsymbol{x}) = V(\boldsymbol{x}) = -\phi(\boldsymbol{x}) + C$, where C is a constant chosen to guarantee the positiveness of $V(\boldsymbol{x})$ over the whole $\mathbb{R}^N$. Note that such constant always exists because of the continuity of ϕ and Assumption 4.4.3. Then, $V(\boldsymbol{x})$ is nonnegative and $V(\boldsymbol{x}) \to \infty$ as $\|\boldsymbol{x}\| \to \infty$ (see again Assumption 4.4.3). Now we show that the function $V(\boldsymbol{x})$ satisfies condition (4.2.2) in Theorem 4.2.2. Using the notation $\tilde{\mathbf{f}}(t, \bar{\boldsymbol{x}}(t)) = \frac{\nabla\phi(\bar{\boldsymbol{x}}(t))}{N} + \boldsymbol{R}(t, \bar{\boldsymbol{x}}(t)) + \boldsymbol{\xi}(t)$ and by the Mean-value Theorem:

$$\begin{aligned} LV(\boldsymbol{x}) &= \mathrm{E}V(\bar{\boldsymbol{x}}(t+1)|\bar{\boldsymbol{x}}(t) = \boldsymbol{x}) - V(\boldsymbol{x}) = \gamma(t+1)\mathrm{E}(\nabla V(\tilde{\boldsymbol{x}}), \tilde{\mathbf{f}}(t, \bar{\boldsymbol{x}}(t))) \\ &= \gamma(t+1)\mathrm{E}(\nabla V(\bar{\boldsymbol{x}}(t)), \tilde{\mathbf{f}}(t, \bar{\boldsymbol{x}}(t))) + \gamma(t+1)\mathrm{E}(\nabla V(\tilde{\boldsymbol{x}}) - \nabla V(\bar{\boldsymbol{x}}(t)), \tilde{\mathbf{f}}(t, \bar{\boldsymbol{x}}(t))), \end{aligned}$$

where $\tilde{\boldsymbol{x}} = \bar{\boldsymbol{x}}(t) - \theta\gamma(t+1)\tilde{\mathbf{f}}(t, \bar{\boldsymbol{x}}(t))$ for some $\theta \in (0,1)$. Taking into account Assumption 4.4.6, we obtain that

$$\|\nabla V(\tilde{\boldsymbol{x}}) - \nabla V(\bar{\boldsymbol{x}}(t))\| \le k_1\gamma(t+1)\|\tilde{\mathbf{f}}(t, \bar{\boldsymbol{x}}(t))\|.$$

Since

$$\begin{aligned} \mathrm{E}(\nabla V(\bar{\boldsymbol{x}}(t)), \tilde{\mathbf{f}}(t, \bar{\boldsymbol{x}}(t))) &= \frac{1}{N}(\nabla V(\bar{\boldsymbol{x}}(t)), \nabla\phi(\bar{\boldsymbol{x}}(t))) + (\nabla V(\bar{\boldsymbol{x}}(t)), \boldsymbol{R}(t, \bar{\boldsymbol{x}}(t))), \\ \mathrm{E}\|\tilde{\mathbf{f}}(t, \bar{\boldsymbol{x}}(t))\|^2 &= \frac{1}{N^2}\|\nabla\phi(\bar{\boldsymbol{x}}(t))\|^2 + \|\boldsymbol{R}(t, \bar{\boldsymbol{x}}(t))\|^2 \\ &\quad + \mathrm{E}\|\boldsymbol{\xi}(t)\|^2 + \frac{2}{N}(\nabla\phi(\bar{\boldsymbol{x}}(t)), \boldsymbol{R}(t, \bar{\boldsymbol{x}}(t))), \end{aligned}$$

and due to the Cauchy–Schwarz inequality and Assumptions 4.4.2, 4.4.7, we conclude that

$$LV(\mathbf{x}) \le \frac{\gamma(t+1)}{N}(\nabla V(\bar{\mathbf{x}}(t)), \nabla\phi(\bar{\mathbf{x}}(t))) + k_2\gamma(t+1)\|\mathbf{R}(t,\bar{\mathbf{x}}(t))\| + k_3\gamma^2(t+1).$$

Now recall that $\nabla V = -\nabla\phi$ and the function $V(\mathbf{x})$ is nonnegative. Thus, we finally obtain that

$$LV(\mathbf{x}) \le -\frac{\gamma(t+1)}{N}\|\nabla\phi(\bar{\mathbf{x}}(t))\|^2 + g(t)(1+V(\bar{\mathbf{x}}(t))),$$

where

$$g(t) = k_2\gamma(t+1)\|\mathbf{R}(t,\bar{\mathbf{x}}(t))\| + k_3\gamma^2(t+1).$$

Lemma 4.4.1 and the choice of $\gamma(t)$ imply that $g(t) > 0$ and $\sum_{t=0}^{\infty} g(t) < \infty$. Thus, all conditions of Theorem 4.2.2 hold and, taking into account $\|\nabla\phi(\bar{\mathbf{x}}(t))\| \in \Psi(\mathbf{A}_0)$, we conclude that $\lim_{t\to\infty}\bar{\mathbf{x}}(t) = \mathbf{a}^*$ almost surely, where $\mathbf{a}^* \in \mathbf{A}_0$ is some critical point of the function ϕ. Next we show that $\mathbf{a}^* \in \mathbf{A}^*$, i.e., the limit point $\mathbf{a}^*$ cannot be a critical point different from a local maximum of ϕ. To do this we use the result formulated in Theorem 4.2.3. Obviously, under Assumptions 4.4.6 and 4.4.7 conditions (4.2.3) and (4.2.3) of Theorem 4.2.3 hold. Thus, it suffices to check finiteness of the sums in conditions (3) and (4) of Theorem 4.2.3. Since $\gamma(t) = O(1/t^{\nu})$, $\nu \in (0.5, 1]$, and

$$\sum_{k=t+1}^{\infty}\gamma^2(k) = \sum_{k=t+1}^{\infty}\frac{1}{k^{2\nu}} \sim \int_{t+1}^{\infty}\frac{1}{x^{2\nu}}dx = O(1/t^{2\nu-1}),$$

we get

$$\sum_{t=0}^{\infty}\left(\frac{\gamma(t)}{\sqrt{\sum_{k=t+1}^{\infty}\gamma^2(k)}}\right)^3 = \sum_{t=0}^{\infty} O(1/t^{1.5}) < \infty.$$

Moreover,

$$\|\mathbf{R}(t,\bar{\mathbf{x}}(t))\| \le c(t) = \frac{L}{N}\sum_{i=1}^{N}\|\mathbf{a}_i(t+1) - \bar{\mathbf{x}}(t)\|$$

almost surely. Hence, according to Theorem 4.3.1 and Assumptions 4.4.2 and 4.4.7,

$$\|\mathbf{R}(t,\bar{\mathbf{x}}(t))\| \le M\left(\lambda^t + \sum_{s=1}^{t}\lambda^{t-s}\gamma(s)\right)$$

almost surely for some positive constant M. Here the parameter $\lambda \in (0, 1)$ is one defined in Theorem 4.3.1. Taking into account the inequality above and the fact that $\frac{\gamma(t)}{\sqrt{\sum_{k=t+1}^{\infty}\gamma^2(k)}} = O(\frac{1}{\sqrt{t}})$, we have

$$\sum_{t=0}^{\infty} \frac{\gamma(t)\|\boldsymbol{R}(t, \bar{\boldsymbol{x}}(t))\|}{\sqrt{\sum_{k=t+1}^{\infty}\gamma^2(k)}} < \infty.$$

The last inequality is obtained due to the fact that, analogously to the proof of Theorem 4.3.4, $\sum_{t=0}^{\infty}\lambda^t O(\frac{1}{\sqrt{t}}) < \infty$ and

$$\sum_{t=0}^{\infty} O\left(\frac{1}{\sqrt{t}}\right)\sum_{s=1}^{t}\lambda^{t-s}\gamma(s) \le \sum_{t=0}^{\infty}\sum_{s=0}^{t}\lambda^{t-s} O\left(\frac{1}{\sqrt{s}}\gamma(s)\right) < \infty.$$

Thus, all conditions of Theorem 4.2.3 are fulfilled. It implies that $\boldsymbol{a}^*$ cannot be a critical point of the function ϕ different of its local maximum, i.e., $\boldsymbol{a}^* \in \boldsymbol{A}^*$. On the other hand, by Theorem 4.3.1, $\lim_{t\to\infty}\|\mathbf{a}_i(t) - \bar{\boldsymbol{x}}(t)\| = 0$ for all $i \in [N]$. It means that $\lim_{t\to\infty}\mathbf{a}_i(t) = \boldsymbol{a}^* \in \boldsymbol{A}^*$ almost surely. Finally, due to the learning algorithm (4.24),

$$a^i(t+1) = \mathrm{a}_i^i(t+1).$$

Hence, the joint action $\boldsymbol{a}(t)$ converges to a local maximum of the potential function $\boldsymbol{a}^*$ almost surely as $t \to \infty$. □

4.4.1 *Simulation Results: Code Division Multiple Access Problem*

Let us consider a regulation problem of a code division multiple access (CDMA) wireless system consisting of three users. Following the framework proposed in the works [ABSA01, SBP06], we model this problem by means of a potential game (see Example 2.2.2 in Sect. 2.2), where users need to maximize the following global objective function:

$$\phi(\boldsymbol{a}) = \log\left(1 + \sum_{i\in[3]} h_i \exp(a^i)\right) - \sum_{i\in[3]} c_i(a^i),$$

where a^i, $i = 1, 2, 3$, is the intensity of the transmitter power of the ith user, $p^i = \exp(a^i)$ is the corresponding transmit power, and the channel gains from the users to the base station are $h_1 = 1$, $h_2 = 0.5$, $h_3 = 1.1$, respectively. More precisely, let the signal to interference ratio of the ith user at the receiver be given by:

$$\mathrm{SINR}_i(\boldsymbol{a}) = \frac{h_i \exp(a^i)}{1 + \sum_{j \neq i} h_j \exp(a^j)}.$$

We assume the cost function of the transmitter power for the ith user to be

$$c_i(a^i) = 3 \log(1 + \exp(a^i)) - a^i.$$

Note that the cost is infinitely large, if the user chooses to transmit with infinite (in absolute value) intensity. This utility is expressed as follows:

$$U_i(\boldsymbol{a}) = \log(1 + \mathrm{SINR}_i(\boldsymbol{a})) - c_i(a^i)$$

and, according to discussion in Example 2.2.2 in Sect. 2.2, results in the potential game $\Gamma = ([3], \boldsymbol{A} = \mathbb{R}^3, \{U_i\}_{i \in [3]}, \phi)$.

To learn a local maximum of the potential function in Γ, the users adapt the communication-based algorithm with the following communication topology. Each two moments of time one of the following two graph combinations is randomly set up for the information exchange:

$$([3], \{1 \leftrightarrow 2, 2 \leftrightarrow 3\})\ ([3], \{1 \leftrightarrow 2, 1 \leftrightarrow 3\}) \text{ or } ([3], \{1 \leftrightarrow 2, 1 \leftrightarrow 3\})\ ([3], \{1 \leftrightarrow 2, 2 \leftrightarrow 3\}).$$

Obviously, such sequence of communication graphs is 4-strongly connected. The step size $\gamma(t)$ is chosen to be $\frac{10}{t^{0.51}}$.

Figures 4.10 and 4.11 demonstrate how the potential function changes with time during the run of the communication-based algorithm with the initial users' estimation vectors $\boldsymbol{x}_1(0) = (1, -1, 4)^T$, $\boldsymbol{x}_2(0) = (-1, 1, 1)^T$, $\boldsymbol{x}_3(0) = (2, -2, 9)^T$. The considerable increase can be noticed already after 50 iterations (see Fig. 4.10). As time runs further the value of ϕ approaches its maximum -4.797 and stays close to it (see Fig. 4.11). Thus, convergence to a local maximum of the potential function $\boldsymbol{a}^* = (-0.33, -0.28, -0.54)$, which is the potential function maximizer, takes place. The algorithm runs analogously, given some other initial estimations (see Figs. 4.12 and 4.13 for the case $\boldsymbol{x}_1(0) = (3, -1, 3)^T$, $\boldsymbol{x}_2(0) = (-2, 3, 3)^T$, $\boldsymbol{x}_3(0) = (5, -3, -4)^T$).

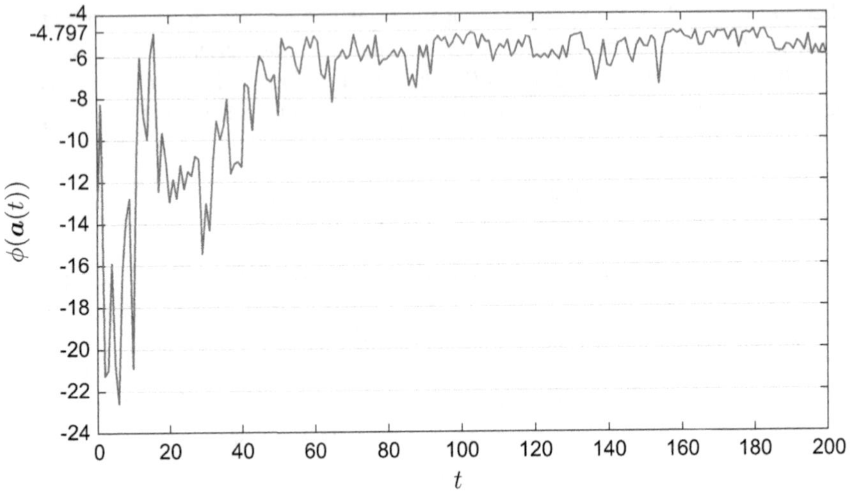

Fig. 4.10 The value of ϕ during the communication-based algorithm (given $\boldsymbol{x}_1(0) = (1, -1, 4)^T$, $\boldsymbol{x}_2(0) = (-1, 1, 1)^T$, $\boldsymbol{x}_3(0) = (2, -2, 9)^T$)

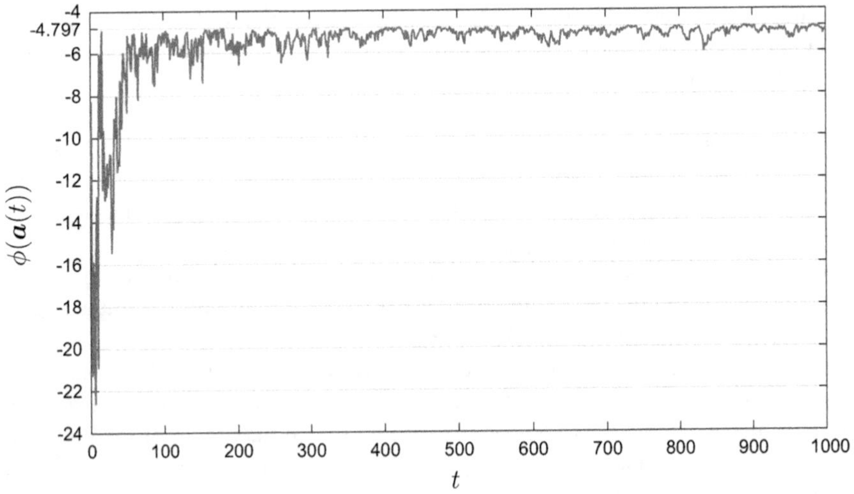

Fig. 4.11 The value of ϕ during the communication-based algorithm (given $\boldsymbol{x}_1(0) = (1, -1, 4)^T$, $\boldsymbol{x}_2(0) = (-1, 1, 1)^T$, $\boldsymbol{x}_3(0) = (2, -2, 9)^T$)

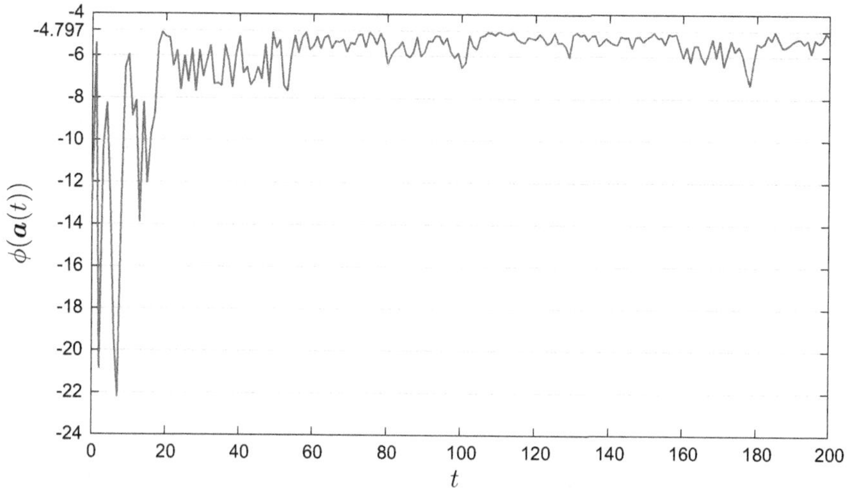

Fig. 4.12 The value of ϕ during the communication-based algorithm (given $\boldsymbol{x}_1(0) = (3, -1, 3)^T$, $\boldsymbol{x}_2(0) = (-2, 3, 3)^T$, $\boldsymbol{x}_3(0) = (5, -3, -4)^T$)

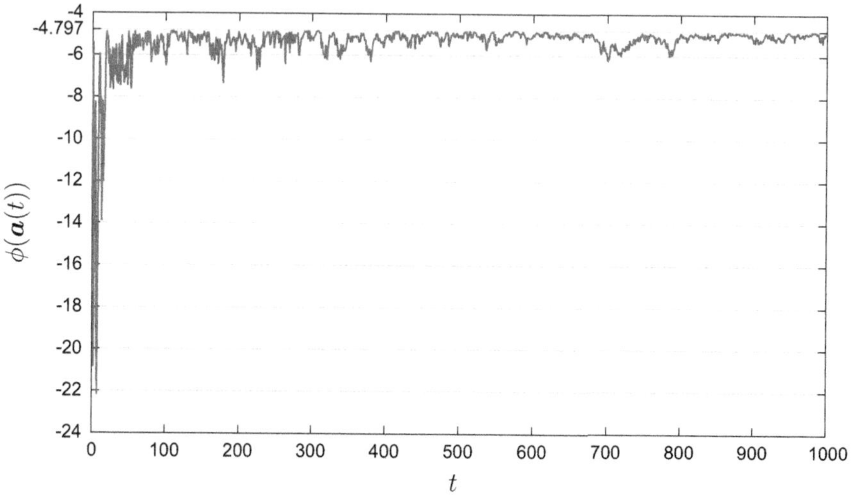

Fig. 4.13 The value of ϕ during the communication-based algorithm (given $\boldsymbol{x}_1(0) = (3, -1, 3)^T$, $\boldsymbol{x}_2(0) = (-2, 3, 3)^T$, $\boldsymbol{x}_3(0) = (5, -3, -4)^T$)

4.5 Payoff-Based Learning in Potential Games

In this section the results on stochastic approximation presented in Sect. 4.2 are used to introduce the *payoff-based algorithm*, where each agent updates her action using only the observation of her played action and the received payoff. Thus, the algorithm is of the following type:

$$a^i(t+1) = F_i(a^i(t), \hat{U}_i(t)), \tag{4.26}$$

where the notation $\hat{U}_i(t) = U_i(\boldsymbol{a}(t))$ is used to emphasize that the agent i does not have an access to the structure of U_i, but she can observe its value at each moment of time t.

As before, the motivation for such learning is the assumption that a given optimization problem in some multi-agent system can be reformulated in terms of learning local maxima of the potential function in a potential game modeled for this system (see Sect. 2.2). In the case of continuous action games, the results on stochastic approximation presented in Sect. 4.2 can be used to develop an algorithm converging in probability to a local maximum of the potential function.

4.5.1 *Convergence to a Local Maximum of the Potential Function*

Analogously to Sect. 4.4, the focus here is on a multi-agent system with N agents, taking actions from $\mathbb{R}$, whose objective is to agree on a joint action $\boldsymbol{a} \in \boldsymbol{A} = \mathbb{R}^N$ corresponding to a *local maximum* of some function $\phi : \mathbb{R}^N \to \mathbb{R}$. Thus, the problem to be solved can be formulated as follows:

$$\begin{aligned} &\max \phi(\boldsymbol{a}), \\ &\text{s.t. } \boldsymbol{a} \in \mathbb{R}^N. \end{aligned} \tag{4.27}$$

According to the discussion in Sect. 2.2, this system is modeled by means of some continuous action potential game $\Gamma = (N, \{A_i\}, \{U_i\}, \phi)$ with $A_i = \mathbb{R}$, $i \in [N]$. Recall that any local solution of (4.27) corresponds to a local maximum of the potential function in the game Γ. As before, the set of local maxima of ϕ is denoted by $\boldsymbol{A}^*$.

The *payoff-based algorithm* is introduced to learn a local maximum of the potential function in the *continuous action* potential game Γ. The learning algorithm updates the agents' mixed strategies, which in their turn are chosen in a way to put more weight on the actions in the direction of the utilities' ascent. We refer here to the idea of CALA (continuous action-set learning automaton), presented in the literature on learning automata [BM06, TS03], and use the Gaussian distribution as the basis for mixing agents' strategies.

Some potential game $\Gamma = (N, \{A_i = \mathbb{R}\}, \{U_i\}, \phi)$ is considered, where each agent i chooses her action from the action set $A_i = \mathbb{R}$ according to the normal distribution $\mathcal{N}(\mu^i, \sigma^2)$ with the density:

$$p_{\mu^i}(x) = \frac{1}{\sqrt{2\pi}\sigma} \exp\left\{-\frac{(x-\mu^i)^2}{2\sigma^2}\right\}. \tag{4.28}$$

The learning algorithm updates the parameter μ^i for every $i \in [N]$ according to the following rule:

$$\mu^i(t+1) = \mu^i(t) + \gamma(t+1)\sigma^3(t+1)\left[\hat{U}_i(t)\frac{a^i(t)-\mu^i(t)}{\sigma^2(t)}\right], \tag{4.29}$$

where $a^i(t)$ and $\hat{U}_i(t)$ are the action played and the payoff experienced at the moment t by the agent i, $\sigma(t)$ is some specified time-dependent variance, and $\gamma(t)$ is a step-size parameter. Obviously, the algorithm above is of type (4.26) and does not require memory (is 1-recall).

For analysis of the payoff-based algorithm above the following assumptions on the potential function will be needed.

Assumption 4.5.1 *The function ϕ in the problem* (4.27) *has finitely many critical points which lie on some compact subset of $\boldsymbol{A} = \mathbb{R}^N$.*

Assumption 4.5.2 *The gradients $\nabla\phi$ exists and is bounded on $\boldsymbol{A} = \mathbb{R}^N$.*

Assumption 4.5.3 *The function $-\phi$ is coercive, i.e., $\phi(\boldsymbol{a}) \to -\infty$ as $\|\boldsymbol{a}\| \to \infty$.*

Assumption 4.5.4 *For any point $\boldsymbol{a}' \in \boldsymbol{A}_0 \setminus \boldsymbol{A}^*$ that is not a local maximum of ϕ there exists a symmetric positive definite matrix $C(\boldsymbol{a}')$ such that $(\nabla\phi(\boldsymbol{a}), C(\boldsymbol{a}')(\boldsymbol{a}-\boldsymbol{a}')) \geq 0$ for any $\boldsymbol{a} \in U(\boldsymbol{a}')$, where $U(\boldsymbol{a}')$ is some open neighborhood of $\boldsymbol{a}'$.*

Assumption 4.5.5 *The gradient $\nabla\phi$ is Lipschitz continuous on $\mathbb{R}^N$, i.e., there exists such a positive constant L that $\|\nabla\phi(\boldsymbol{a}_1) - \nabla\phi(\boldsymbol{a}_2)\| \leq L\|\boldsymbol{a}_1 - \boldsymbol{a}_2\|$ for any $\boldsymbol{a}_1, \boldsymbol{a}_2 \in \mathbb{R}^N$.*

Remark 4.5.1 Since Assumption 4.5.2 requires the boundedness of $\nabla\phi$, the function $-\phi(\boldsymbol{a})$ grows not faster than a linear function as $\boldsymbol{a} \to \infty$, given fulfillment of Assumptions 4.5.2 and 4.5.3. Moreover, in this case, one can assume existence of some constant $K > 0$ such that $a^i \frac{\partial\phi(\boldsymbol{a})}{\partial a^i} \leq 0$ for any $\boldsymbol{a} : \|\boldsymbol{a}\| > K$.

Moreover, the following assumption regarding the utility functions is made.

Assumption 4.5.6 *The utility functions $U_i(\boldsymbol{a})$, $i \in [N]$, are continuous on the whole $\mathbb{R}^N$ and their absolute values grow not faster than a linear function of $\boldsymbol{a}$, given $\boldsymbol{a}$ has a large norm, i.e., there exist such constants $K > 0$ and $c > 0$ that $|U_i(\boldsymbol{a})| \leq c(1 + \|\boldsymbol{a}\|)$ for any $i \in [N]$, if $\|\boldsymbol{a}\| \geq K$.*

The following sections provide the analysis of the algorithm (4.29). It will be shown that under Assumptions 4.5.1–4.5.6 on the potential function and an appropriate choice of the parameters $\sigma(t)$ and $\gamma(t)$, the mixed joint actions chosen according to $\prod_{i=1}^{N} \mathcal{N}(\mu^i(t), \sigma(t))$, where $\mu^i(t)$, $i \in [N]$, are updated as in (4.29), converge in probability to a *pure* joint action coinciding with a local maximum of the potential function as time tends to infinity.[9]

Firstly, we calculate the mathematical expectation of the term $\hat{U}_i(a^i - \mu^i)$ given that the coordinates of the joint action $\boldsymbol{a}$ are independently distributed and the ith coordinate has the distribution $\mathcal{N}(\mu^i, \sigma^2)$.

$$\mathrm{E}\{\hat{U}_i(a^i - \mu^i)|a^i \sim \mathcal{N}(\mu^i, \sigma), i \in [N]\}$$

$$= \int_{\mathbb{R}^N} U_i(\boldsymbol{x})(x^i - \mu^i) p_{\mu^1}(x^1) \cdot \ldots \cdot p_{\mu^N}(x^N) d\boldsymbol{x}, \tag{4.30}$$

where $\boldsymbol{x} = (x^1, \ldots, x^N)$ and p_{μ^i}, $i \in [N]$, is introduced in (4.28). On the other hand, the utility function $\tilde{U}_i(\boldsymbol{\mu})$, $\boldsymbol{\mu} = (\mu^1, \ldots, \mu^N)$, in the mixed strategy game is

$$\tilde{U}_i(\boldsymbol{\mu}) = \int_{\mathbb{R}^N} U_i(\boldsymbol{x}) p_{\mu^1}(x^1) \cdot \ldots \cdot p_{\mu^N}(x^N) d\boldsymbol{x}.$$

Under fulfillment of Assumption 4.5.2, we can differentiate the expression above under the integral sign and, thus, obtain[10]

$$\frac{\partial \tilde{U}_i(\boldsymbol{\mu})}{\partial \mu^i} = \int_{\mathbb{R}^N} U_i(\boldsymbol{x}) p_{\mu^1}(x^1) \cdot \ldots \frac{\partial p_{\mu^i}(x^i)}{\partial \mu^i} \ldots \cdot p_{\mu^N}(x^N) d\boldsymbol{x}$$

$$= \int_{\mathbb{R}^N} \frac{(x^i - \mu^i)}{\sigma^2} U_i(\boldsymbol{x}) p_{\mu^1}(x^1) \cdot \ldots \cdot p_{\mu^N}(x^N) d\boldsymbol{x}. \tag{4.31}$$

Bringing (4.30) and (4.31) together, we get

$$\mathrm{E}\{\hat{U}_i(a^i - \mu^i)|a^i \sim \mathcal{N}(\mu^i, \sigma), i \in [N]\} = \sigma^2 \frac{\partial \tilde{U}_i(\boldsymbol{\mu})}{\partial \mu^i}.$$

Taking this equality and properties of mixed strategy potential games (see Proposition 2.1.2 and equality (2.3) in Sect. 2.1.3) into account, we conclude that $\frac{\partial \tilde{U}_i(\boldsymbol{\mu})}{\partial \mu^i} = \frac{\partial \tilde{\phi}(\boldsymbol{\mu})}{\partial \mu^i}$ for all $i \in [N]$. Thus, the algorithm (4.29) can be rewritten as follows:

$$\boldsymbol{\mu}(t+1) = \boldsymbol{\mu}(t) + \sigma^3(t+1)\gamma(t+1)[\nabla\phi(\boldsymbol{\mu}(t)) + \boldsymbol{Q}(\boldsymbol{\mu}(t))]$$

$$+ \gamma(t+1)[\sigma^3(t+1)\boldsymbol{M}(t, \boldsymbol{\mu}(t))], \tag{4.32}$$

---

[9] $\prod_{i=1}^{N} \mathcal{N}(\mu^i(t), \sigma(t))$ is the N-dimensional normal distribution with independently distributed coordinates, where each coordinate $i \in [N]$ has the distribution $\mathcal{N}(\mu^i(t), \sigma(t))$.

[10] For more details on explanation of differentiation under the integral sign see Sect. 4.5.2.

where $\boldsymbol{Q}(\boldsymbol{\mu}(t)) = \nabla\tilde{\phi}(\boldsymbol{\mu}(t)) - \nabla\phi(\boldsymbol{\mu}(t))$ and $\boldsymbol{M}(t, \boldsymbol{\mu}(t))$ is the N-dimensional vector with the coordinates

$$M_i(t, \boldsymbol{\mu}(t)) = \hat{U}_i(t)\frac{a^i(t) - \mu^i(t)}{\sigma^2(t)} - \frac{\partial \tilde{U}_i(\boldsymbol{\mu})}{\partial \mu^i}, \quad i \in [N].$$

Thus,

$$\mathrm{E}\{\boldsymbol{M}(t, \boldsymbol{\mu}(t))|a^i \sim \mathcal{N}(\mu^i(t), \sigma(t)), i \in [N]\} = \mathbf{0}. \tag{4.33}$$

Now the main result on the payoff-based procedure (4.29) and its vector form (4.32) can be formulated.

Theorem 4.5.1 *Let* $\Gamma = (N, \{A_i\}, \{U_i\}, \phi)$ *be a* continuous *action potential game with* $A_i = \mathbb{R}$. *Let the parameters* $\gamma(t)$ *and* $\sigma(t)$ *be such that* $\gamma(t) > 0$, $\sigma(t) > 0$,

(1) $\sum_{t=0}^{\infty} \gamma(t)\sigma^3(t) = \infty$, $\sum_{t=0}^{\infty} \gamma(t)\sigma^4(t) < \infty$,

(2) $\sum_{t=0}^{\infty} \gamma^2(t) < \infty$, $\sum_{t=0}^{\infty} \left(\frac{\gamma(t)}{\sqrt{\sum_{k=t+1}^{\infty} \gamma^2(k)}}\right)^3 < \infty$,

(3) $\sum_{t=0}^{\infty} \frac{\gamma(t)\sigma^4(t)}{\sqrt{\sum_{k=t+1}^{\infty} \gamma^2(k)}} < \infty$.

Then under Assumptions 4.5.1–4.5.6 the sequence $\{\boldsymbol{\mu}(t)\}$ *defined in* (4.32) *converges to some* $\boldsymbol{\mu}^* \in \boldsymbol{A}^*$ *almost surely. Moreover, the payoff-based algorithm defined in* (4.29) *converges in probability to a local maximum* $\boldsymbol{a}^* = \boldsymbol{\mu}^*$ *of the potential function* ϕ.

Remark 4.5.2 There exist such sequences $\{\gamma(t)\}$, $\{\sigma(t)\}$ that conditions (4.5.1)–(4.5.1) of Theorem 4.5.1 hold. For example, let us consider $\gamma(t) = 1/t^{0.6}$ and $\sigma(t) = 1/t^{0.13}$. Notice that in this case

$$\sum_{k=t+1}^{\infty} \gamma^2(k) = \sum_{k=t+1}^{\infty} \frac{1}{k^{1.2}} \sim \int_{t+1}^{\infty} \frac{1}{x^{1.2}} dx = O(1/t^{0.2}).$$

Hence,

$$\sum_{t=0}^{\infty} \left(\frac{\gamma(t)}{\sqrt{\sum_{k=t+1}^{\infty} \gamma^2(k)}}\right)^3 = \sum_{t=0}^{\infty} O(1/t^{1.5}) < \infty,$$

$$\sum_{t=0}^{\infty} \frac{\gamma(t)\sigma^4(t)}{\sqrt{\sum_{k=t+1}^{\infty} \gamma^2(k)}} = \sum_{t=0}^{\infty} O(1/t^{1.02}) < \infty.$$

Proof We will deal with the equivalent formulation (4.32) of the learning algorithm (4.29). In the following any k_j denotes some positive constant. To prove the claim we will demonstrate that the process (4.32) fulfills the conditions in Theorems 4.2.2 and 4.2.3, where $d = N$, $\mathbf{X}(t) = \boldsymbol{\mu}(t)$, $\alpha(t) = \gamma(t)\sigma^3(t)$, $\beta(t) = \gamma(t)$, $\mathbf{f}(\mathbf{X}(t)) = \nabla\phi(\boldsymbol{\mu})$, $\mathbf{q}(t, \mathbf{X}(t)) = \boldsymbol{Q}(\boldsymbol{\mu}(t))$, and $\mathbf{W}(t, \mathbf{X}(t), \omega) = \sigma^3(t)\boldsymbol{M}(t, \boldsymbol{\mu}(t))$. First of all, we show that there exists a sample function $V(t, \boldsymbol{\mu})$ of the process (4.32) satisfying conditions (4.2.2) and (4.2.2) of Theorem 4.2.2. Let us consider the following time-invariant function:

$$V(\boldsymbol{\mu}) = -\phi(\boldsymbol{\mu}) + \sum_{i=1}^{N} h(\mu^i) + C,$$

where[11]

$$h(\mu^i) = \begin{cases} (\mu^i - K)^2, & \text{if } \mu_i \geq K; \\ 0, & \text{if } |\mu_i| \leq K; \\ (\mu^i + K)^2, & \text{if } \mu_i \leq -K; \end{cases}$$

and C is chosen in such way that $V(\boldsymbol{\mu}) > 0$ for all $\boldsymbol{\mu} \in \mathbb{R}^N$. Thus, $V(\boldsymbol{\mu})$ is positive on $\mathbb{R}^N$ and $\lim_{\|\boldsymbol{\mu}\| \to \infty} V(\boldsymbol{\mu}) = \infty$. We can use the Mean-value Theorem to get

$$\begin{aligned} LV(\boldsymbol{\mu}) &= \mathrm{E}V(\boldsymbol{\mu}(t+1)|\boldsymbol{\mu}(t) = \boldsymbol{\mu}) - V(\boldsymbol{\mu}) \\ &= \mathrm{E}V(\boldsymbol{\mu} + \sigma^3(t+1)\gamma(t+1)\tilde{\mathbf{f}}(t, \boldsymbol{\mu})) - V(\boldsymbol{\mu}) \\ &= \mathrm{E}\sigma^3(t+1)\gamma(t+1)(\nabla V(\tilde{\boldsymbol{\mu}}), \tilde{\mathbf{f}}(t, \boldsymbol{\mu})) \\ &= \sigma^3(t+1)\gamma(t+1)\mathrm{E}\{(\nabla V(\boldsymbol{\mu}), \tilde{\mathbf{f}}(t, \boldsymbol{\mu})) \\ &\quad + (\nabla V(\tilde{\boldsymbol{\mu}}) - \nabla V(\boldsymbol{\mu}), \tilde{\mathbf{f}}(t, \boldsymbol{\mu}))\}, \end{aligned} \tag{4.34}$$

where

$$\begin{aligned} \tilde{\mathbf{f}}(t, \boldsymbol{\mu}) &= \nabla\phi(\boldsymbol{\mu}) + \boldsymbol{Q}(\boldsymbol{\mu}) + \boldsymbol{M}(t, \boldsymbol{\mu}), \\ \tilde{\boldsymbol{\mu}} &= \boldsymbol{\mu} + \theta\sigma^3(t+1)\gamma(t+1)\tilde{\mathbf{f}}(t, \boldsymbol{\mu}) \end{aligned}$$

for some $\theta \in (0, 1)$. We proceed by estimating the terms in (4.34). Let $h(\boldsymbol{\mu}) = \sum_{i=1}^{N} h_i(\boldsymbol{\mu})$. Then, taking into account (4.33) and the fact that the vector $\nabla h(\boldsymbol{\mu})$ has coordinates that are linear in $\boldsymbol{\mu}$, we get:

$$\begin{aligned} \mathrm{E}\{(\nabla V(\boldsymbol{\mu}), \tilde{\mathbf{f}}(t, \boldsymbol{\mu}))\} &= -(\|\nabla\phi(\boldsymbol{\mu})\|^2 - (\nabla\phi(\boldsymbol{\mu}), \nabla h(\boldsymbol{\mu}))) + (\nabla h(\boldsymbol{\mu}) - \nabla\phi(\boldsymbol{\mu}), \boldsymbol{Q}(\boldsymbol{\mu})) \\ &\leq -(\|\nabla\phi(\boldsymbol{\mu})\|^2 - (\nabla\phi(\boldsymbol{\mu}), \nabla h(\boldsymbol{\mu}))) + k_1\|\boldsymbol{Q}(\boldsymbol{\mu})\|(1 + V(\boldsymbol{\mu})), \end{aligned}$$

[11]The constant K below is one from Remark 4.5.1.

where the last inequality is due to the Cauchy–Schwarz inequality and Assumption 4.5.2. Thus, using again the Cauchy–Schwarz inequality, Assumption 4.5.5, and the fact that $\nabla V(\boldsymbol{\mu})$ is Lipschitz continuous, we obtain from (4.34):

$$\begin{aligned} LV(\boldsymbol{\mu},t) &\le -\sigma^3(t+1)\gamma(t+1)(\|\nabla\phi(\boldsymbol{\mu})\|^2 - (\nabla\phi(\boldsymbol{\mu}),\nabla h(\boldsymbol{\mu}))) \\ &\quad + \sigma^3(t+1)\gamma(t+1)k_2[\|\boldsymbol{Q}(\boldsymbol{\mu})\| \\ &\quad + \sigma^3(t+1)\gamma(t+1)\mathrm{E}\|\tilde{\mathbf{f}}(t,\boldsymbol{\mu}))\|^2](1+V(\boldsymbol{\mu})). \end{aligned} \tag{4.35}$$

Next, taking into account Assumptions 4.5.2, 4.5.3, and Remark 4.5.1, we can write[12] for any fixed $\boldsymbol{\mu}$:

$$\begin{aligned} \frac{\partial\tilde{\phi}(\boldsymbol{\mu})}{\partial\mu^i} &= \frac{1}{\sigma^2}\int_{\mathbb{R}^N}\phi(\boldsymbol{x})(x^i-\mu^i)p_{\mu}(\boldsymbol{x})d\boldsymbol{x} = -\int_{\mathbb{R}^N}\phi(\boldsymbol{x})p_{\mu^{-i}}(\boldsymbol{x}^{-i})d\left(e^{-\frac{(x^i-\mu^i)^2}{2\sigma^2}}\right)d\boldsymbol{x}^{-i} \\ &= -\int_{\mathbb{R}^{N-1}}\left(\phi(\boldsymbol{x})e^{-\frac{(x^i-\mu^i)^2}{2\sigma^2}}\right)\Bigg|_{-\infty}^{\infty}p_{\mu^{-i}}(\boldsymbol{x}^{-i})d\boldsymbol{x}^{-i} + \int_{\mathbb{R}^N}\frac{\partial\phi(\boldsymbol{x})}{\partial x^i}p_{\mu}(\boldsymbol{x})d\boldsymbol{x} \\ &= \int_{\mathbb{R}^N}\frac{\partial\phi(\boldsymbol{x})}{\partial x^i}p_{\mu}(\boldsymbol{x})d\boldsymbol{x}, \end{aligned} \tag{4.36}$$

where $p_{\mu}(\boldsymbol{x}) = \prod_{i=1}^N p_{\mu^i}(x^i)$ with p_{μ^i} defined in (4.28), and $p_{\mu^{-i}}(\boldsymbol{x}^{-i}) = \prod_{j\neq i}p_{\mu^j}(x^j)$. Thus,

$$\nabla\tilde{\phi}(\boldsymbol{\mu}) = \int_{\mathbb{R}^N}\nabla\phi(\boldsymbol{x})p_{\mu}(\boldsymbol{x})d\boldsymbol{x}. \tag{4.37}$$

Since $\boldsymbol{Q}(\boldsymbol{\mu}(t)) = \nabla\tilde{\phi}(\boldsymbol{\mu}(t)) - \nabla\phi(\boldsymbol{\mu}(t))$ and due to Assumptions 4.5.2, 4.5.5 and (4.37), we can write the following:

$$\begin{aligned} \|\boldsymbol{Q}(\boldsymbol{\mu}(t))\| &\le \int_{\mathbb{R}^N}\|\nabla\phi(\boldsymbol{\mu})-\nabla\phi(\boldsymbol{x})\|p_{\mu}(\boldsymbol{x})d\boldsymbol{x} \le \int_{\mathbb{R}^N}L\|\boldsymbol{\mu}-\boldsymbol{x}\|p_{\mu}(\boldsymbol{x})d\boldsymbol{x} \\ &\le \int_{\mathbb{R}^N}L\left(\sum_{i=1}^N|\mu^i-x^i|\right)p_{\mu}(\boldsymbol{x})d\boldsymbol{x} = O(\sigma(t)), \end{aligned} \tag{4.38}$$

Due to Assumption 4.5.6 and (4.30), $\sigma^6(t+1)\mathrm{E}\|\boldsymbol{M}(t,\boldsymbol{\mu})\|^2$ is bounded by a quadratic function of $\boldsymbol{\mu}$. Hence, $\sigma^6(t+1)\mathrm{E}\|\tilde{\mathbf{f}}(t,\boldsymbol{\mu}))\|^2 \le k_3(1+V(\boldsymbol{\mu}))$, and (4.35) implies:

$$LV(\boldsymbol{\mu},t) \le -\sigma^3(t+1)\gamma(t+1)(\|\nabla\phi(\boldsymbol{\mu})\|^2 - (\nabla\phi(\boldsymbol{\mu}),\nabla h(\boldsymbol{\mu}))) + g(t)(1+V(\boldsymbol{\mu})), \tag{4.39}$$

[12]For more details on explanation of differentiation under the integral sign we refer the reader to Sect. 4.5.2.

where $g(t) = O(\sigma^4(t)\gamma(t) + \gamma^2(t))$, i.e., $\sum_{t=1}^{\infty} g(t) < \infty$, according to the choice of the sequence $\gamma(t)$ and $\sigma(t)$ (see conditions (4.5.1) and (4.5.1)). Note also that, according to the definition of the function h, $\|\nabla\phi(\boldsymbol{\mu})\|^2 - (\nabla\phi(\boldsymbol{\mu}), \nabla h(\boldsymbol{\mu})) \geq 0$, where equality holds only on critical points of the function ϕ. Thus, conditions (4.2.2) and (4.2.2) of Theorem 4.2.2 hold. Conditions (4.2.2) and (4.2.2) of Theorem 4.2.2 hold, due to (4.38) and Assumption 4.5.2, respectively. Moreover, taking into account Theorem 4.2.1 and (4.39), we conclude that the norm $\|\boldsymbol{\mu}(t)\|$ is bounded almost surely for all t. Hence, condition (4.2.2) of Theorem 4.2.2 holds as well. Thus, all conditions of Theorem 4.2.2 are fulfilled. It implies that $\lim_{t\to\infty} \boldsymbol{\mu}(t) = \boldsymbol{\mu}^*$ almost surely, where $\boldsymbol{\mu}^*$ is some critical point of ϕ: $\nabla\phi(\boldsymbol{\mu}^*) = 0$. Moreover, since $\sigma(t) \to 0$ as $t \to \infty$, we conclude that $\boldsymbol{a}(t)$ converges to $\boldsymbol{\mu}^*$ in probability and the learning algorithm under consideration converges to some critical point $\boldsymbol{a}^* = \boldsymbol{\mu}^*$ of the potential function ϕ in probability.

Further we verify the fulfillment of conditions in Theorem 4.2.3 to prove that this critical point $\boldsymbol{a}^* = \boldsymbol{\mu}^*$ can be nothing but a local maximum of ϕ. Let $\boldsymbol{\mu}'$ denote a critical point of the function ϕ that is not in the set of local maxima $\boldsymbol{A}^*$.

We show that there exists some $\delta' > 0$ such that $|\mathrm{Tr}[A(t,\boldsymbol{\mu}) - A(t,\boldsymbol{\mu}')]| \leq k_4\|\boldsymbol{\mu} - \boldsymbol{\mu}'\|$, for any $\boldsymbol{\mu}$: $\|\boldsymbol{\mu} - \boldsymbol{\mu}'\| < \delta'$, where $A_{ii}(t,\boldsymbol{\mu}) = \mathrm{E}\sigma^6(t+1)M_i^2(t,\boldsymbol{\mu})$. Indeed,

$$\frac{1}{\sigma^6(t+1)}|\mathrm{Tr}[A(t,\boldsymbol{\mu}) - A(t,\boldsymbol{\mu}')]| = \left|\sum_{i=1}^{N} \mathrm{E}\{M_i^2(t,\boldsymbol{\mu}) - M_i^2(t,\boldsymbol{\mu}')\}\right|$$

$$\leq \left|\sum_{i=1}^{N} \mathrm{E}\left[\left\{\hat{U}_i^2(t)\frac{(a_i(t)-\mu^i)^2}{\sigma^4(t)}\right|\right.\right.$$

$$\left. a_i(t) \sim \mathcal{N}(\mu^i,\sigma), i \in [N]\right\}$$

$$\left.\left. - \left\{\hat{U}_i^2(t)\frac{(a_i(t)-\mu_i')^2}{\sigma^4(t)}\right| a_i(t) \sim \mathcal{N}(\mu_i',\sigma), i \in [N]\right\}\right]\right|$$

$$+ \|\nabla\tilde{\phi}(\boldsymbol{\mu}')\|^2 - \|\nabla\tilde{\phi}(\boldsymbol{\mu})\|^2. \tag{4.40}$$

Since $\nabla\phi$ is Lipschitz continuous, we can use (4.37) to get

$$\|\nabla\tilde{\phi}(\boldsymbol{\mu}) - \nabla\tilde{\phi}(\boldsymbol{\mu}')\| = \left\|\int_{\mathbb{R}^N} (\nabla\phi(\boldsymbol{x})p_{\mu}(\boldsymbol{x}) - \nabla\phi(\boldsymbol{x})p_{\mu'}(\boldsymbol{x}))d\boldsymbol{x}\right\|$$

$$\leq \int_{\mathbb{R}^N} \|\nabla\phi(y+\boldsymbol{\mu}) - \nabla\phi(y+\boldsymbol{\mu}')\|p(y)dy \leq k_5\|\boldsymbol{\mu} - \boldsymbol{\mu}'\|,$$

where $p(y) = \frac{1}{(2\pi\sigma^2)^{N/2}}e^{-\frac{Ny^2}{2\sigma^2}}$. Hence, according to Assumptions 4.5.2, 4.5.5 and the last inequality,

$$\|\nabla\tilde{\phi}(\boldsymbol{\mu}')\|^2 - \|\nabla\tilde{\phi}(\boldsymbol{\mu})\|^2 \leq k_6(\|\nabla\tilde{\phi}(\boldsymbol{\mu}')\| - \|\nabla\tilde{\phi}(\boldsymbol{\mu})\|)$$

$$\leq k_6\|\nabla\tilde{\phi}(\boldsymbol{\mu}') - \nabla\tilde{\phi}(\boldsymbol{\mu})\| \leq k_7\|\boldsymbol{\mu} - \boldsymbol{\mu}'\|. \tag{4.41}$$

Moreover,

$$\sigma^6\mathrm{E}\left[\left\{\hat{U}_i^2\frac{(a_i-\mu^i)^2}{\sigma^4}|a_i\sim\mathcal{N}(\mu^i,\sigma),i\in[N]\right\}-\sigma^6\left\{\hat{U}_i^2\frac{(a_i-\mu_i')^2}{\sigma^4}|a_i\sim\mathcal{N}(\mu_i',\sigma),i\in[N]\right\}\right]$$
$$=u_i(t,\boldsymbol{\mu})-u_i(t,\boldsymbol{\mu}'), \tag{4.42}$$

where

$$u_i(\boldsymbol{\mu})=\int_{\mathbb{R}^N}\sigma^2U_i^2(\boldsymbol{x})(x^i-\mu^i)^2p_{\mu}(\boldsymbol{x})d\boldsymbol{x}. \tag{4.43}$$

The function above is Lipschitz continuous in some δ-neighborhood of $\boldsymbol{\mu}'$, since there exists $\delta>0$ such that the gradient $\nabla u_i(\boldsymbol{\mu})$ is bounded for any $\boldsymbol{\mu}$: $\|\boldsymbol{\mu}-\boldsymbol{\mu}'\|<\delta$. Indeed, due to Assumption 4.5.1 and if $\|\boldsymbol{\mu}-\boldsymbol{\mu}'\|<\delta$, the mean vectors $\boldsymbol{\mu}'$ and $\boldsymbol{\mu}$ in (4.42) are bounded. Next, taking into account linear behavior of each $U_i(\boldsymbol{x})$, when $\boldsymbol{x}\to\infty$ (see Assumption 4.5.6), and finiteness of moments of a normal random vector for a bounded mean vector $\boldsymbol{\mu}$, we can apply the sufficient condition for uniform convergence of integrals based on majorants [ZC04] to conclude that the integral in (4.43) can be differentiated under the integral sign with respect to the parameter μ^j for any $j\in[N]$. For the same reason of moments' finiteness, the partial derivative $\frac{\partial u_i(\boldsymbol{\mu})}{\partial\mu^j}$ is bounded for all $i,j\in[N]$. According to the Mean-value Theorem, this implies that

$$\sum_{i=1}^N|u_i(\boldsymbol{\mu})-u_i(\boldsymbol{\mu}')|\le k_8\|\boldsymbol{\mu}-\boldsymbol{\mu}'\|. \tag{4.44}$$

Substituting (4.41)–(4.44) into (4.40) and taking into account that, according to Assumption 4.5.5, $\|\nabla\phi(\boldsymbol{\mu})\|^2=\|\nabla\phi(\boldsymbol{\mu})-\nabla\phi(\boldsymbol{\mu}')\|^2\le L^2\|\boldsymbol{\mu}-\boldsymbol{\mu}'\|^2$ for all $\boldsymbol{\mu}\in\mathbb{R}^N$, we obtain that there exists $\delta'\le\delta$ such that for any $\boldsymbol{\mu}:\|\boldsymbol{\mu}-\boldsymbol{\mu}'\|<\delta'$

$$\|\nabla\phi(\boldsymbol{\mu})\|^2+|\mathrm{Tr}[A(t,\boldsymbol{\mu})-A(t,\boldsymbol{\mu}')]|\le k_8\|\boldsymbol{\mu}-\boldsymbol{\mu}'\|.$$

Thus, condition (4.2.3) of Theorem 4.2.3 holds.

Since $\|\boldsymbol{\mu}'\|<\infty$ (Assumption 4.5.1) and due to Assumptions 4.5.2, 4.5.3, and Remark 4.5.1, $\sigma^{12}\mathrm{E}\|\boldsymbol{M}(t,\boldsymbol{\mu})\|^4<\infty$ for any $\boldsymbol{\mu}:\|\boldsymbol{\mu}-\boldsymbol{\mu}'\|<\delta'$. Hence, condition (4.2.3) of Theorem 4.2.3 is also fulfilled.

Finally, taking into account the choice of the sequences $\{\gamma(t)\}$, $\{\sigma(t)\}$ (see conditions (4.5.1)–(4.5.1)) and the estimation (4.38), we conclude that the last two conditions of Theorem 4.2.3 are also fulfilled. It allows us to conclude that the almost sure limit $\boldsymbol{\mu}^*$ of the process (4.32) cannot be equal to $\boldsymbol{\mu}'\in\boldsymbol{A}_0\setminus\boldsymbol{A}^*$, but $\boldsymbol{\mu}^*\in\boldsymbol{A}^*$. As $\lim_{t\to0}\sigma(t)=0$, the players' joint actions $\boldsymbol{a}(t)$, chosen according to the rules in (4.29), converge in probability (see Theorem A.1.2 in Appendix A.1) to a local maximum $\boldsymbol{a}^*=\boldsymbol{\mu}^*$ of the potential function ϕ. □

4.5.2 Additional Calculations

Here we justify the differentiation under the integral sign in the expressions (4.31) and (4.36) under Assumption 4.5.2. Obviously, the functions under the integral sign, namely $U_i(\boldsymbol{x})(x^i - \mu^i)p_{\mu}(\boldsymbol{x})$ and $\phi(\boldsymbol{x})(x^i - \mu^i)p_{\mu}(\boldsymbol{x})$, respectively, are continuous given Assumption 4.5.2. Thus, it remains to check that the integrals of these functions over the whole $\mathbb{R}^N$ converge uniformly in respect to $\mu^i \in \mathbb{R}$. We can write using the Taylor expansion of the function ϕ around the point $\boldsymbol{\mu}$:

$$\begin{aligned}\int_{\mathbb{R}^N} \phi(\boldsymbol{x})(x^i - \mu^i)p_{\mu}(\boldsymbol{x})d\boldsymbol{x} &= \int_{\mathbb{R}^N} (\phi(\boldsymbol{\mu}) + (\nabla\phi(\boldsymbol{\eta}(\boldsymbol{x},\boldsymbol{\mu})), \boldsymbol{x} - \boldsymbol{\mu}))(x^i - \mu^i)p_{\mu}(\boldsymbol{x})d\boldsymbol{x} \\ &= \int_{\mathbb{R}^N} (\nabla\phi(\boldsymbol{\eta}(\boldsymbol{x},\boldsymbol{\mu})), \boldsymbol{x} - \boldsymbol{\mu})(x^i - \mu^i)p_{\mu}(\boldsymbol{x})d\boldsymbol{x} \\ &= \int_{\mathbb{R}^N} (\nabla\phi(\tilde{\boldsymbol{\eta}}(\boldsymbol{y},\boldsymbol{\mu})), \boldsymbol{y})y^i p(\boldsymbol{y})d\boldsymbol{y},\end{aligned}$$

where $\boldsymbol{\eta}(\boldsymbol{x},\boldsymbol{\mu}) = \boldsymbol{\mu} + \theta(\boldsymbol{x} - \boldsymbol{\mu})$, $\theta \in (0,1)$, $\boldsymbol{y} = \boldsymbol{x} - \boldsymbol{\mu}$, $\tilde{\boldsymbol{\eta}}(\boldsymbol{y},\boldsymbol{\mu}) = \boldsymbol{\mu} + \theta\boldsymbol{y}$, $p(\boldsymbol{y}) = \frac{1}{(2\pi\sigma^2)^{N/2}} e^{-\frac{Ny^2}{2\sigma^2}}$. The uniform convergence of the integral above follows from the fact that, under Assumption 4.5.2, $\|\nabla\phi(\tilde{\boldsymbol{\eta}}(\boldsymbol{y},\boldsymbol{\mu}))\| \le C$ for some positive constant C and, hence,

$$|(\nabla\phi(\tilde{\boldsymbol{\eta}}(\boldsymbol{y},\boldsymbol{\mu})), \boldsymbol{y})y_i p(\boldsymbol{y})| \le h(\boldsymbol{y}) = C\|\boldsymbol{y}\|y^i p(\boldsymbol{y}),$$

where $\int_{\mathbb{R}^N} h(\boldsymbol{y})d\boldsymbol{y} < \infty$ (see the basic sufficient condition using majorant [ZC04]). The argumentation for the function $U_i(\boldsymbol{x})(x^i - \mu^i)p_{\mu}(\boldsymbol{x})$ is similar.

4.5.3 Simulation Results: Power Transition Problem

Here the same setting of optimal control of power transition in CDMA as in Sect. 4.4.1 is considered. Now it is assumed that each user can observe only her currently played action and experienced payoff. Thus, the payoff-based algorithm should be used to learn a local optimum in such situation. Recall that there are three users, whose local utilities are a tradeoff between achieving a certain signal to interference ratio and minimizing costs of transmitter power. The signal to interference plus noise ratio of the ith user at the receiver is given by:

$$\mathrm{SINR}_i(\boldsymbol{a}) = \frac{h_i \exp(a^i)}{1 + \sum_{j\neq i} h_j \exp(a^j)},$$

where a^i, $i = 1, 2, 3$, is the intensity of the transmitter power of the ith user, $p^i = \exp(a^i)$ is the corresponding transmit power, and $h_1 = 1$, $h_2 = 0.5$, $h_3 = 1.1$ are the channel gains from the users to the base station.

The utility function of the user i is expressed as follows:

$$U_i(\boldsymbol{a}) = \log(1 + \mathrm{SINR}_i(\boldsymbol{a})) - c_i(a^i),$$

where c_i is the cost function of the transmitter power for the ith user:

$$c_i(a^i) = 3\log(1 + \exp(a^i)) - a^i.$$

The game $\Gamma = ([3], \boldsymbol{A} = \mathbb{R}^3, \{U_i\}_{i\in[3]}, \phi)$ is potential with the potential function

$$\phi(\boldsymbol{a}) = \log\left(1 + \sum_{i\in[3]} h_i \exp(a^i)\right) - \sum_{i\in[3]} c_i(a^i),$$

which coincides with the system's objective one.

Figures 4.14 and 4.15 demonstrate the performance of the algorithm after 200 and 1000, iterations respectively, given the following initial values of $\boldsymbol{\mu}(0)$: $\mu^1(0) = 5$, $\mu^2(0) = 3$, $\mu^3(0) = -1$. We can see that the potential function gets close to its maximum -4.797 already after 200 steps of the algorithm and approximates this value further on. If the initial value of $\boldsymbol{\mu}(0)$ is changed to $\mu^1(0) = 0$, $\mu^2(0) = -20$, $\mu^3(0) = -10$, the algorithm's run changes insignificantly (see Figs. 4.16 and 4.17). Since $\boldsymbol{\mu}(0) = (0, -20, -10)$ corresponds to a point that

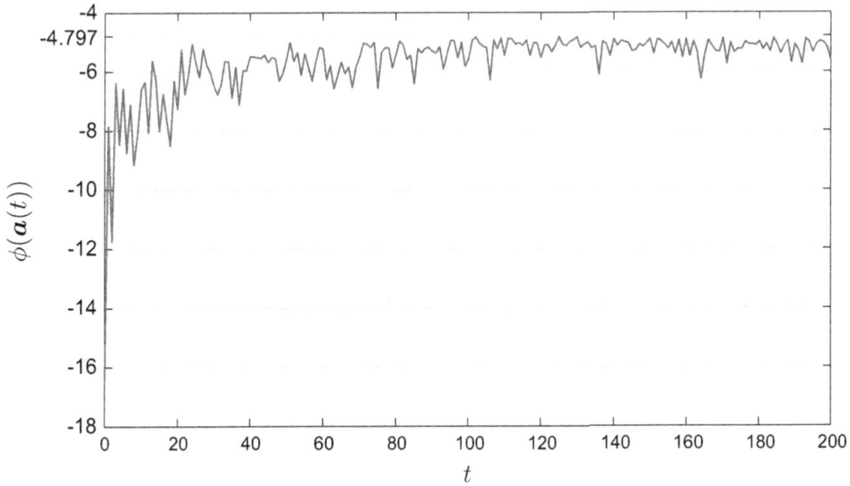

Fig. 4.14 The value of ϕ during the payoff-based learning algorithm (given $\mu^1(0) = 5$, $\mu^2(0) = 3$, $\mu^3(0) = -1$)

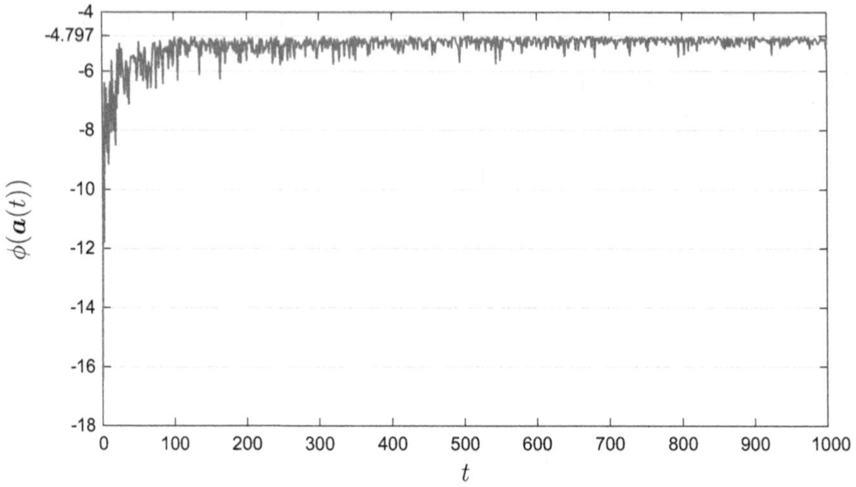

Fig. 4.15 The value of ϕ during the payoff-based learning algorithm (given $\mu^1(0) = 5$, $\mu^2(0) = 3$, $\mu^3(0) = -1$)

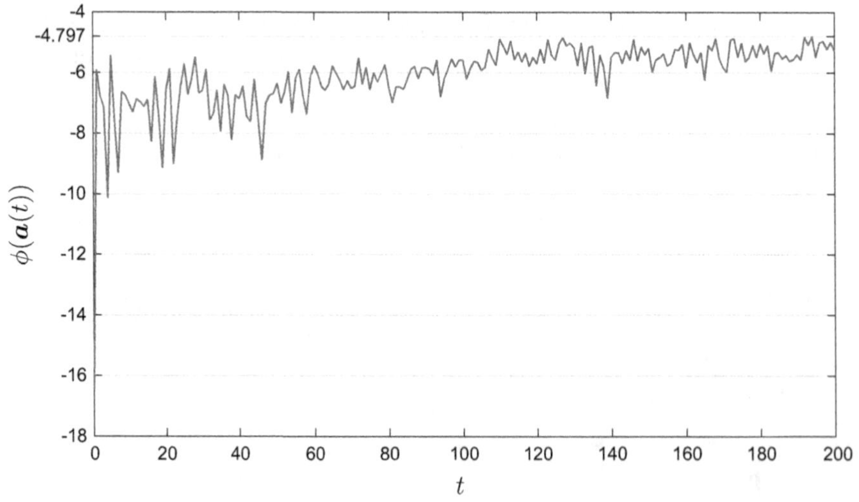

Fig. 4.16 The value of ϕ during the payoff-based learning algorithm (given $\mu^1(0) = 0$, $\mu^2(0) = -20$, $\mu^3(0) = -10$)

is more distant from the closest local optimum $\boldsymbol{a}^* = (-0.33, -0.28, -0.54)$ in comparison to $\boldsymbol{\mu}(0) = (5, 3, -1)$, the algorithm needs more time to approximate $\boldsymbol{a}^*$ and, thus, to guarantee a high value of ϕ (see Fig. 4.16). However, after sufficient approximation reached in 200 steps the value of ϕ stays close to -4.797 (see Fig. 4.17).

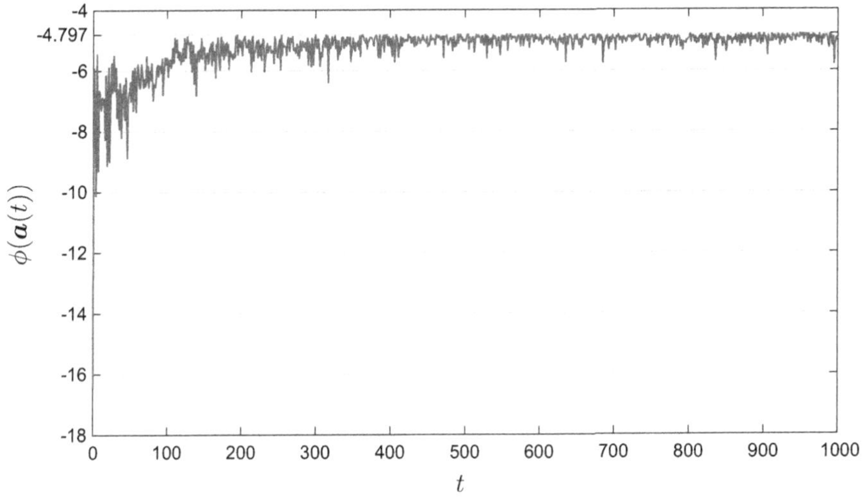

Fig. 4.17 The value of ϕ during the payoff-based learning algorithm (given $\mu^1(0) = 0, \mu^2(0) = -20, \mu^3(0) = -10$)

4.6 Concave Potential Games with Uncoupled Action Sets

In the previous section the payoff-based algorithm was formulated for potential games with unconstrained action sets. This section demonstrates how this algorithm can be adapted to deal with potential games with convex, bounded action sets.

4.6.1 *Problem Formulation and Assumptions*

The focus here is on a potential game $\Gamma = (N, \{A_i\}, \{U_i\}, \phi)$ with $A_i \subset \mathbb{R}^d$ for all $i \in [N]$.[13] As before, this game is modeled to find a maximum of the potential function $\phi : \boldsymbol{A} \to \mathbb{R}$, where $\boldsymbol{A} = A_1 \times \cdots \times A_N$. This maximum coincides, on one hand, with a Nash equilibrium in Γ and, on the other hand, is equal to an optimal state of the corresponding multi-agent system. In contrast to Sect. 4.5, the following assumptions regarding the game Γ are made.

Assumption 4.6.1 *The potential game under consideration is concave. Namely, the set A_i is convex and compact for all $i \in [N]$ and the potential function is concave on $\boldsymbol{A}$.*

[13]All results below are applicable for games with different dimensions $\{d_i\}$ of the action sets $\{A_i\}$.

Assumption 4.6.2 *The gradient of the function ϕ exists on the set $\mathbf{A}$ and $\nabla\phi$ is a Lipschitz function on $\mathbf{A}$.*

Assumption 4.6.3 *The functions ϕ and U_i, $i \in [N]$, can be defined outside the joint action set $\mathbf{A}$ as differentiable functions such that $\nabla\phi$ is a Lipschitz function on $\mathbb{R}^{Nd}$ and $|U_i|$ grows not faster than a polynomial function as the argument tends to infinity for any $i \in [N]$.*

Thus, the problem under consideration can be formulated as follows:

$$\begin{aligned} &\phi(\boldsymbol{a}) \to \max, \\ &\text{s.t. } \boldsymbol{a} \in \boldsymbol{A}. \end{aligned} \tag{4.45}$$

Note that, according to Assumption 4.6.1, the problem above is equivalent to the problem of finding a Nash equilibrium in the game Γ. Moreover, the problem (4.45) has a non-empty solution (see Theorem 2.1.1).

To motivate consideration of such problems, the following subsection presents a system of an electricity market, where users intend to find a stable point over their consumption strategies.

4.6.2 Applications to Electricity Markets

For more details on game models in electricity markets we refer the reader to [Jen10, TGM14]. There are N market participants (users), also referred to as players. Let $\boldsymbol{a}_i = [a_i^1, \dots, a_i^d] \in \mathbb{R}^d$ denote the decision variable of the player i, $i = 1, \dots, N$, that is the vector corresponding to her consumption profile over d periods. The constraints for each player i are

$$\begin{aligned} &0 \le a_i^k \le \bar{a} \quad \text{for } k = 1, \dots, d, \\ &\sum_{k=1}^{d} a_i^k = \bar{a}_i. \end{aligned} \tag{4.46}$$

They indicate that for each player the electricity consumption at each time instance is limited and the total electricity consumption over the considered period of time needs to match a desired amount. The convex and compact set defined by the constraints in (4.46) is considered the action set A_i for the corresponding player i.

The cost function is the price paid for electricity consumption by each agent [Jen10, PKL16]

$$J_i(\boldsymbol{a}_i, \boldsymbol{a}_{-i}) = \boldsymbol{a}_i^\top Q \boldsymbol{a}_i + 2\left(C\frac{1}{N}\sum_{j=1}^{N} \boldsymbol{a}_j + \mathbf{c}\right)^\top \boldsymbol{a}_i \tag{4.47}$$

with $Q, C \in \mathbb{R}^{d\times d}$, $\mathbf{c} \in \mathbb{R}^d$.

Consider the following setup. At iteration t, each player submits its proposed consumption profile over time $\mathbf{x}_i(t) = [x_i^1(t), \dots, x_i^d(t)]$.[14] Then she receives the value of the cost $\hat{J}_i(t) = J_i(\mathbf{x}_i(t), \mathbf{x}_{-i}(t))$. This corresponds to the total electricity price she would need to pay over time interval $[1, \dots, d]$ for this choice. It is naturally to assume that each agent knows her action set A_i.

How should players update their profiles in order to make the sequence of the joint profiles $\{\mathbf{x}(t) = (\mathbf{x}_i(t), \mathbf{x}_{-i}(t))\}_t$ *convergent to a Nash equilibrium in the game* $\Gamma = (N, \{A_i\}, \{U_i = -J_i\})$ *as* $t \to \infty$*?*

The matrices Q and C are assumed to be symmetric and, moreover, $Q \succeq 0$ and $C \succeq 0$. Under such setting the game $\Gamma = (N, \{A_i\}_i, \{U_i = -J_i\}_i, \phi)$ is potential with the potential function $\phi = -\psi$, where $\psi : \mathbb{R}^{Nd} \to \mathbb{R}$ is expressed by

$$\psi(\mathbf{x}) = \frac{1}{2}\mathbf{x}^\top A\mathbf{x} + \mathbf{b}^\top \mathbf{x}, \tag{4.48}$$

$A = I_N \otimes \left(Q + Q^\top + \frac{2C^\top}{N}\right) + (\mathbb{1}_N \mathbb{1}_N^\top) \otimes \frac{2C}{N}$, $\mathbf{b} = \mathbb{1}_N \otimes 2\mathbf{c}$, $\mathbf{x} = [\mathbf{x}_1, \dots, \mathbf{x}_N]^\top$ [Jen10]. According to the assumptions on the matrices Q and C, the function $\psi(\mathbf{x})$ is convex on $\mathbb{R}^{Nd}$. This implies existence of a Nash equilibrium in Γ coinciding with a maximizer of ϕ (minimizer of ψ) over the set of joint actions $\boldsymbol{A} = A_1 \times \dots \times A_N$ that is compact and convex. Thus, the problem of finding a Nash equilibrium in Γ can be reformulated as follows:

$$\begin{aligned} &\phi(\boldsymbol{a}) \to \max, \\ &\text{s.t. } \boldsymbol{a} \in \boldsymbol{A} \subset \mathbb{R}^{Nd}. \end{aligned} \tag{4.49}$$

Since the cost functions are quadratic and the matrices Q and C are positive semidefinite, all Assumptions 4.6.1–4.6.3 are fulfilled for the formulated problem (4.49), given $U_i = -J_i$ for all i.

4.6.3 *Payoff-Based Algorithm in Concave Games*

This subsection formulates the payoff-based approach for the distributed solution of the problem (4.45) resulting in learning a Nash equilibrium $\boldsymbol{a}_{\mathrm{NE}}$ in a potential game $\Gamma = (N, \{A_i\}, \{U_i\}, \phi)$ satisfying Assumptions 4.6.1–4.6.3. Having access to the information about the current state $\mathbf{x}_i(t)$ at the iteration t and the current utility value $\hat{U}_i(t)$ at the joint state $\mathbf{x}(t)$, $\hat{U}_i(t) = U_i(\mathbf{x}_1(t), \dots, \mathbf{x}_N(t))$, each agent

[14] Note that this profile must not belong to the action set A_i defined by the constraints in (4.47).

chooses her next state[15] $\mathbf{x}_i(t+1)$ randomly according to the multidimensional normal distribution $\mathcal{N}(\boldsymbol{\mu}_i(t+1), \sigma(t+1))$ with the density:

$$p_i(x_i^1, \ldots, x_i^d; \boldsymbol{\mu}_i(t+1), \sigma(t+1)) = \frac{1}{(\sqrt{2\pi}\sigma)^d} \exp\left\{-\sum_{k=1}^{d} \frac{(x_i^k - \mu_i^k(t+1))^2}{2\sigma^2(t+1)}\right\},$$

where the parameter $\boldsymbol{\mu}_i(t)$ is updated as follows:

$$\boldsymbol{\mu}_i(t+1) = \mathrm{Proj}_{A_i}\left[\boldsymbol{\mu}_i(t) + \gamma(t+1)\sigma^2(t+1)\hat{U}_i(t)\frac{\mathbf{x}_i(t) - \boldsymbol{\mu}_i(t)}{\sigma^2(t)}\right], \tag{4.50}$$

where $\mathrm{Proj}_C[\cdot]$ denotes the operator of projection on some set C. Taking into account[16]

$$\mathrm{E}\left\{\hat{U}_i(t)\frac{x_i^k(t) - \mu_i^k(t)}{\sigma^2(t)} | x_i^k(t) \sim \mathcal{N}(\mu_i^k(t), \sigma(t))\right\} = \frac{\partial \tilde{J}_i(\boldsymbol{\mu}_1(t), \ldots, \boldsymbol{\mu}_N(t), \sigma(t))}{\partial \mu_i^k},$$

for any $i \in [N]$, $k \in [d]$, where

$$\tilde{U}_i(\boldsymbol{\mu}_1, \ldots, \boldsymbol{\mu}_N, \sigma) = \int_{\mathbb{R}^{Nd}} J_i(\boldsymbol{x}) p_{\boldsymbol{\mu}}(\boldsymbol{x}) d\boldsymbol{x}, \; p_{\boldsymbol{\mu}}(\boldsymbol{x}) = \prod_{j=1}^{N} p_j(x_j^1, \ldots, x_j^d; \boldsymbol{\mu}_j, \sigma),$$

we obtain the following vector form for the algorithm:

$$\begin{aligned}\boldsymbol{\mu}(t+1) = \mathrm{Proj}_{\boldsymbol{A}}[&\boldsymbol{\mu}(t) + \gamma(t+1)\sigma^2(t+1) \\ &\times(\nabla\phi(\boldsymbol{\mu}(t)) + \boldsymbol{Q}(\boldsymbol{\mu}(t), \sigma(t)) + \boldsymbol{R}(\boldsymbol{\mu}(t), \sigma(t)))],\end{aligned} \tag{4.51}$$

where

$$\begin{aligned}\boldsymbol{Q}(\boldsymbol{\mu}(t), \sigma(t)) &= \nabla\tilde{\phi}(\boldsymbol{\mu}(t)) - \nabla\phi(\boldsymbol{\mu}(t)), \\ \boldsymbol{R}(\boldsymbol{\mu}(t), \sigma(t)) &= \boldsymbol{F}(t, \mathbf{x}(t), \boldsymbol{\mu}(t)) - \nabla\tilde{\phi}(\boldsymbol{\mu}(t)), \\ \boldsymbol{F}(t, \mathbf{x}(t), \boldsymbol{\mu}(t)) &= \left(\hat{U}_1(t)\frac{\mathbf{x}_1(t) - \boldsymbol{\mu}_1(t)}{\sigma^2(t)}, \ldots, \hat{U}_N(t)\frac{\mathbf{x}_N(t) - \boldsymbol{\mu}_N(t)}{\sigma^2(t)}\right)^\top, \\ \tilde{\phi}(\boldsymbol{\mu}(t)) &= \int_{\mathbb{R}^{Nd}} \phi(\boldsymbol{x}) p_{\boldsymbol{\mu}(t)}(\boldsymbol{x}) d\boldsymbol{x}.\end{aligned}$$

---

[15]We distinguish here between states and actions, since the joint state $\mathbf{x}(t+1) = (\mathbf{x}_1(t+1), \ldots, \mathbf{x}_N(t+1))$ updated during the payoff-based algorithm under consideration must not belong to the set $\boldsymbol{A}$.

[16]Explanation is similar to one for the equality (4.31).

To demonstrate efficient behavior of the procedure above, we will use further the following theorem that is a well-known result of Robbins and Siegmund on nonnegative random variables (see, for example, Lemma 10 in [Pol87]).

Theorem 4.6.1 *Let (Ω, F, P) be a probability space and $F_1 \subset F_2 \subset \ldots$ a sequence of sub-σ-algebras of F. Let z_n, β_n, ξ_n, and ζ_n be nonnegative F_n-measurable random variables such that*

$$\mathrm{E}(z_{n+1}|F_n) \le z_n(1+\beta_n) + \xi_n - \zeta_n.$$

Then almost surely $\lim_{n\to\infty} z_n$ exists and is finite. Moreover, $\sum_{n=1}^{\infty} \zeta_n < \infty$ almost surely on $\{\sum_{n=1}^{\infty} \beta_n < \infty, \sum_{n=1}^{\infty} \xi_n < \infty\}$.

Let $\beta(t+1) = \gamma(t+1)\sigma^2(t+1)$. Then[17]

$$\boldsymbol{\mu}(t+1) = \mathrm{Proj}_{\boldsymbol{A}}[\boldsymbol{\mu}(t) + \beta(t+1)(\nabla\phi(\boldsymbol{\mu}(t)) + \boldsymbol{Q}(\boldsymbol{\mu}(t)) + \boldsymbol{R}(\boldsymbol{\mu}(t)))]. \tag{4.52}$$

Let $\boldsymbol{\mu}^* \in \mathbb{R}^{Nd}$ be a solution of the problem (4.45). Then, taking into account the iterative procedure for the update of $\boldsymbol{\mu}(t)$ above and the non-expansion property of the projecting operator on a convex set, we obtain

$$\begin{aligned}
&\|\boldsymbol{\mu}(t+1) - \boldsymbol{\mu}^*\|^2 \\
&\quad = \|\mathrm{Proj}_{\boldsymbol{A}}[\boldsymbol{\mu}(t) + \beta(t+1)(\nabla\phi(\boldsymbol{\mu}(t)) + \boldsymbol{Q}(\boldsymbol{\mu}(t)) + \boldsymbol{R}(\boldsymbol{\mu}(t)))] - \boldsymbol{\mu}^*\|^2 \\
&\quad \le \|\boldsymbol{\mu}(t) + \beta(t+1)(\nabla\phi(\boldsymbol{\mu}(t)) + \boldsymbol{Q}(\boldsymbol{\mu}(t)) + \boldsymbol{R}(\boldsymbol{\mu}(t))) - \boldsymbol{\mu}^*\|^2 \\
&\quad = \|\boldsymbol{\mu}(t) - \boldsymbol{\mu}^*\|^2 + 2\beta(t+1)(\nabla\phi(\boldsymbol{\mu}(t)) + \boldsymbol{Q}(\boldsymbol{\mu}(t)) + \boldsymbol{R}(\boldsymbol{\mu}(t)), \boldsymbol{\mu}(t) - \boldsymbol{\mu}^*) \\
&\qquad + \beta^2(t+1)\|\nabla\phi(\boldsymbol{\mu}(t)) + \boldsymbol{Q}(\boldsymbol{\mu}(t)) + \boldsymbol{R}(\boldsymbol{\mu}(t))\|^2.
\end{aligned}$$

Taking into account feasibility of $\boldsymbol{\mu}(t)$, $\boldsymbol{\mu}(t) \in \boldsymbol{A}$ for any t, and convexity of ϕ (Assumption 4.6.1), we conclude that for any t

$$(\nabla\phi(\boldsymbol{\mu}(t)), \boldsymbol{\mu}(t) - \boldsymbol{\mu}^*) \le \phi(\boldsymbol{\mu}(t)) - \phi(\boldsymbol{\mu}^*) \le 0.$$

Hence,

$$\begin{aligned}
\mathrm{E}\{\|\boldsymbol{\mu}(t+1) - \boldsymbol{\mu}^*\|^2|\mathcal{F}_t\} \le{}& \|\boldsymbol{\mu}(t) - \boldsymbol{\mu}^*\|^2 - 2\beta(t+1)(\phi(\boldsymbol{\mu}^*) - \phi(\boldsymbol{\mu}(t))) \\
&+ 2\beta(t+1)\|\boldsymbol{Q}(\boldsymbol{\mu}(t))\|\|\boldsymbol{\mu}(t) - \boldsymbol{\mu}^*\| \\
&+ \beta^2(t+1)\mathrm{E}\|g(\boldsymbol{\mu}(t))\|^2,
\end{aligned} \tag{4.53}$$

[17] We omit further the argument $\sigma(t)$ in terms $\boldsymbol{Q}$ and $\boldsymbol{R}$ for the sake of notation simplicity.

where $g(\boldsymbol{\mu}(t)) = \nabla\phi(\boldsymbol{\mu}(t)) + \boldsymbol{Q}(\boldsymbol{\mu}(t)) + \boldsymbol{R}(\boldsymbol{\mu}(t))$ and $\mathcal{F}_t$ is the σ-algebra generated by the random variables $\{\boldsymbol{\mu}(k), k \leq t\}$.

Note that, according to Assumption 4.6.2,[18]

$$\|\boldsymbol{Q}(\boldsymbol{\mu}(t))\| \leq O(\sigma(t)). \tag{4.54}$$

Obviously,

$$\|\boldsymbol{\mu}(t) - \boldsymbol{\mu}^*\| \leq (1 + \|\boldsymbol{\mu}(t) - \boldsymbol{\mu}^*\|^2). \tag{4.55}$$

Now we proceed estimating the term $\mathrm{E}\|g(\boldsymbol{\mu}(t))\|^2$:

$$\begin{aligned} \mathrm{E}\|g(\boldsymbol{\mu}(t))\|^2 \leq \|\nabla\phi(\boldsymbol{\mu}(t))\|^2 + \|\boldsymbol{Q}(\boldsymbol{\mu}(t))\|^2 + 2\|\nabla\phi(\boldsymbol{\mu}(t))\|\|\boldsymbol{Q}(\boldsymbol{\mu}(t))\| \\ + \mathrm{E}\|\boldsymbol{R}(\boldsymbol{\mu}(t))\|^2. \end{aligned} \tag{4.56}$$

Note that

$$\mathrm{E}\|\boldsymbol{R}(\boldsymbol{\mu}(t))\|^2 \leq \sum_{i=1}^{N}\sum_{k=1}^{d}\int_{\mathbb{R}^{Nd}} U_i^2(\boldsymbol{x})\frac{(x_i^k - \mu_i^k(t))^2}{\sigma^4(t)} p_{\boldsymbol{\mu}(t)}(\boldsymbol{x})d\boldsymbol{x}.$$

Thus, we can use Assumption 4.6.3 to conclude that

$$\mathrm{E}\|\boldsymbol{R}(\boldsymbol{\mu}(t))\|^2 \leq \frac{1}{\sigma^4(t)} f(\boldsymbol{\mu}(t), \sigma(t)),$$

where $f(\boldsymbol{\mu}(t), \sigma(t))$ is a polynomial of $\boldsymbol{\mu}(t)$ and $\sigma(t)$. Hence, taking into account boundedness of $\boldsymbol{\mu}(t)$ for all t and Assumption 4.6.2, we conclude that

$$\beta^2(t+1)\mathrm{E}\|\boldsymbol{R}(\boldsymbol{\mu}(t))\|^2 \leq k_3\gamma^2(t) \tag{4.57}$$

for some constant k_3. Moreover, according to boundedness of $\boldsymbol{\mu}(t)$ for all t, we can conclude that the first term on the right-hand side of (4.56) is bounded.

Bringing (4.53)–(4.57) together, we obtain

$$\begin{aligned} \mathrm{E}\{\|\boldsymbol{\mu}(t+1) - \boldsymbol{\mu}^*\|^2 | \mathcal{F}_t\} \leq \|\boldsymbol{\mu}(t) - \boldsymbol{\mu}^*\|^2 - 2\beta(t+1)(\phi(\boldsymbol{\mu}^*) - \phi(\boldsymbol{\mu}(t))) \\ + k_4\beta(t+1)\sigma(t)(1 + \|\boldsymbol{\mu}(t) - \boldsymbol{\mu}^*\|^2) \\ + k_2\beta^2(t+1)(1 + \|\boldsymbol{\mu}(t) - \boldsymbol{\mu}^*\|^2) + (\beta(t+1)\sigma(t))^2 \\ + k_3\gamma^2(t+1) + k_1\beta^2(t+1)\sigma(t)(1 + \|\boldsymbol{\mu}(t) - \boldsymbol{\mu}^*\|^2). \end{aligned}$$

[18]Explanation is similar to one for the equality (4.38).

Thus,

$$\begin{aligned}\mathrm{E}\{\|\boldsymbol{\mu}(t+1)-\boldsymbol{\mu}^*\|^2|\mathcal{F}_t\} \leq & \|\boldsymbol{\mu}(t)-\boldsymbol{\mu}^*\|^2(1+h(t)) \\ & -2\beta(t+1)(\phi(\boldsymbol{\mu}^*)-\phi(\boldsymbol{\mu}(t))) \\ & +h(t)+(\beta(t+1)\sigma(t))^2+k_3\gamma^2(t+1), \quad (4.58)\end{aligned}$$

where $h(t) = k_4\beta(t+1)\sigma(t) + k_2\beta^2(t+1) + k_1\beta^2(t+1)\sigma(t)$.

Now the main result of this section can be formulated.

Theorem 4.6.2 *Let the players in a* continuous *action potential game* Γ *update their states* $\{\mathbf{x}_i(t+1)\}$ *at time* $t+1$ *according to the normal distribution* $\mathcal{N}(\boldsymbol{\mu}_i(t+1), \sigma(t+1))$, *where the mean parameters* $\{\boldsymbol{\mu}_i(t+1)\}$ *are updated as in* (4.50)–(4.51). *Let Assumptions 4.6.1–4.6.3 hold and the variance parameter* $\sigma(t)$ *and the step-size parameter* $\gamma(t)$ *be chosen such that* $\sum_{t=0}^{\infty}\gamma(t)\sigma^2(t) = \infty$, $\sum_{t=0}^{\infty}\gamma(t)\sigma^3(t) < \infty$, *and* $\sum_{t=0}^{\infty}\gamma^2(t) < \infty$. *Then, as* $t \to \infty$, *the mean vector* $\boldsymbol{\mu}(t) = (\boldsymbol{\mu}_1(t), \ldots, \boldsymbol{\mu}_N(t))$ *converges almost surely to a solution of the problem* (4.45), *and the joint state* $\mathbf{x}(t)$ *converges in probability to a pure joint action that is a Nash equilibrium of the game* Γ.

Proof According to (4.58) and the choice of the parameters $\sigma(t)$ and $\gamma(t)$,

$$\begin{aligned}\mathrm{E}\{\|\boldsymbol{\mu}(t+1)-\boldsymbol{\mu}^*\|^2|\mathcal{F}_t\} \leq & \|\boldsymbol{\mu}(t)-\boldsymbol{\mu}^*\|^2(1+h(t)) \\ & -2\gamma(t+1)\sigma^2(t+1)(\phi(\boldsymbol{\mu}^*) \\ & -\phi(\boldsymbol{\mu}(t)))+h_1(t), \quad (4.59)\end{aligned}$$

where $\sum_{t=0}^{\infty} h(t) < \infty$ and $\sum_{t=0}^{\infty} h_1(t) < \infty$. Thus, we can use the result formulated in Theorem 4.6.1 to conclude that almost surely

$$\|\boldsymbol{\mu}(t+1)-\boldsymbol{\mu}^*\|^2 \text{ has a finite limit} \quad (4.60)$$

and

$$\sum_{t=0}^{\infty}\gamma(t+1)\sigma^2(t+1)(\phi(\boldsymbol{\mu}^*)-\phi(\boldsymbol{\mu}(t))) < \infty.$$

Since $\sum_{t=0}^{\infty}\gamma(t)\sigma^2(t) = \infty$, the last convergent series implies existence of the subsequent $\boldsymbol{\mu}(t_l)$ such that almost surely

$$\lim_{l\to\infty}\phi(\boldsymbol{\mu}(t_l)) = \phi(\boldsymbol{\mu}^*). \quad (4.61)$$

Moreover, as $\|\boldsymbol{\mu}(t+1)-\boldsymbol{\mu}^*\|^2$ converges almost surely, $\boldsymbol{\mu}(t_l)$ is bounded almost surely and, hence, there exists an almost surely convergent subsequence $\boldsymbol{\mu}(t_{l_s})$ of the sequence $\boldsymbol{\mu}(t_l)$. According to (4.61),

$$\lim_{s\to\infty}\phi(\boldsymbol{\mu}(t_{l_s})) = \phi(\boldsymbol{\mu}^*) \quad \text{almost surely.}$$

This implies that $\lim_{s\to\infty} \boldsymbol{\mu}(t_{l_s}) = \hat{\boldsymbol{\mu}}$ almost surely, where $\phi(\hat{\boldsymbol{\mu}}) = \phi(\boldsymbol{\mu}^*)$. Taking into account (4.60) and since the choice of $\boldsymbol{\mu}^*$ as a solution of the problem (4.45) is not specified, we conclude that almost surely $\|\boldsymbol{\mu}(t+1) - \hat{\boldsymbol{\mu}}\|^2$ has a finite limit. Thus, almost surely

$$\lim_{t\to\infty} \boldsymbol{\mu}(t) = \hat{\boldsymbol{\mu}},$$

where $\hat{\boldsymbol{\mu}}$ is a solution of the problem (4.45). As $\lim_{t\to\infty} \sigma(t) = 0$, the last assertion of theorem follows as well. □

4.6.4 Simulation Results: Electricity Market Application

Let us consider an example of an electricity market introduced in Sect. 4.6.2. Let us assume that there are N users, whose strategies are their consumption profiles for $d = 4$ periods, the matrices Q and C in their cost functions (4.47) are the identity matrices of the size $4N \times 4N$, and the vector $\mathbf{c}$ is a four-dimensional vector, whose coordinates are some random variables taking values in the interval $(0, 5)$. We assume that the action set A_i for each user $i \in [N]$ is defined by (4.46), where $\bar{a} = 6$ and $\bar{a}_i$ is a random variable taking values in the interval $(0.5, 10)$.

Let the agents follow the payoff-based algorithm described by (4.50)–(4.51). Figures 4.18, 4.19, 4.20, and 4.21 present behavior of the norm $\|\boldsymbol{\mu}(t) - \mathbf{x}_{\mathrm{NE}}\|$

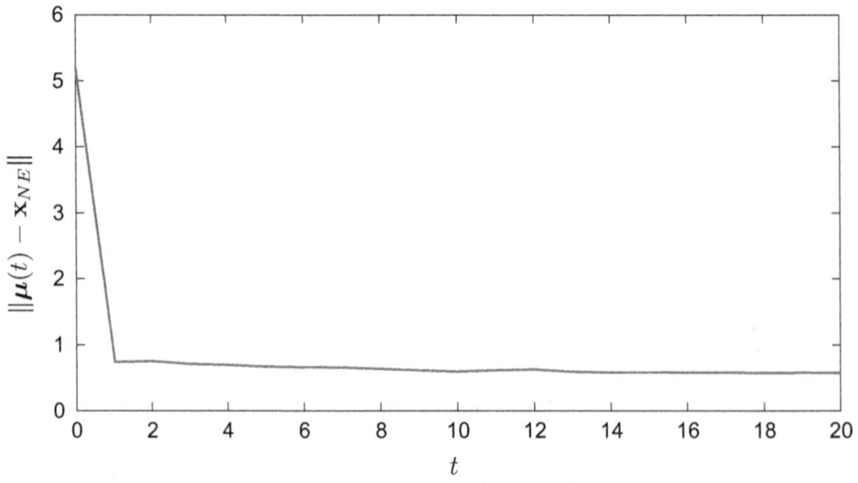

Fig. 4.18 $\|\boldsymbol{\mu}(t) - \mathbf{x}_{\mathrm{NE}}\|$ during the payoff-based algorithm, $N = 10$, $\gamma(t) = \frac{1}{t^{0.51}}$, $\sigma(t) = \frac{0.1}{t^{0.2}}$

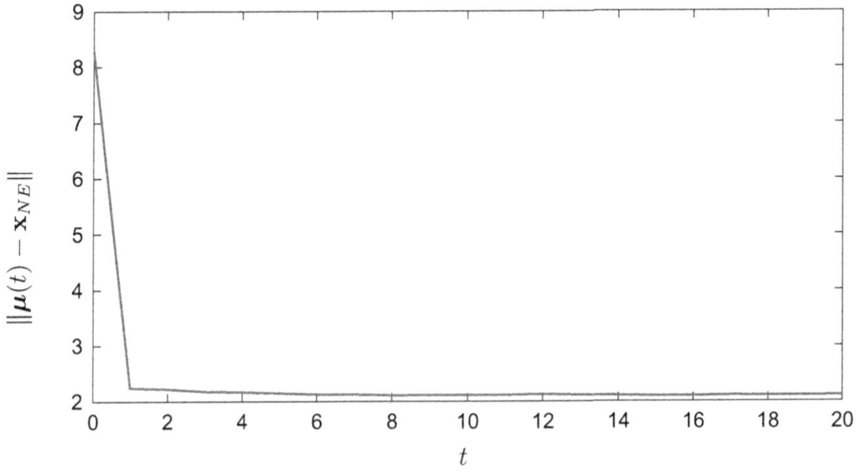

Fig. 4.19 $\|\mu(t) - \mathbf{x}_{\rm NE}\|$ during the payoff-based algorithm, $N = 100$, $\gamma(t) = \frac{1}{t^{0.51}}$, $\sigma(t) = \frac{0.1}{t^{0.2}}$

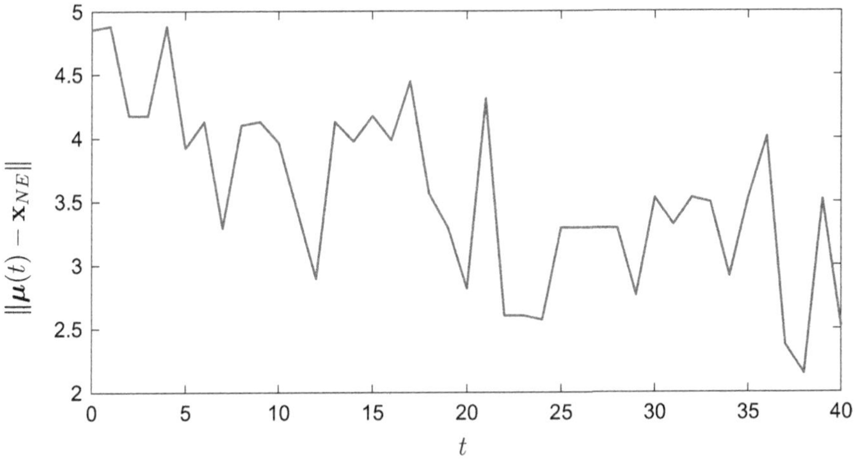

Fig. 4.20 $\|\mu(t) - \mathbf{x}_{\rm NE}\|$ during the payoff-based algorithm, $N = 100$, $\gamma(t) = \frac{1}{t^{0.6}}$, $\sigma(t) = \frac{1}{t^{0.1}}$

during the algorithm's run, where $\mathbf{x}_{\rm NE}$ is the unique[19] Nash equilibrium of the corresponding potential game estimated as the maximum of the potential function over the joint strategy set. In the case of Figs. 4.18 and 4.19, $\gamma(t) = \frac{1}{t^{0.51}}$, $\sigma(t) = \frac{0.1}{t^{0.2}}$, $N = 10$, and $N = 100$, respectively. We can see that already after the first iteration the algorithm gives an approximation for the Nash equilibrium in the game. At the next iterations the value of the vector $\mu(t)$ approaches $\mathbf{x}_{\rm NE}$ slowly. This can be

[19]The uniqueness is due to the fact that the cost functions are strictly convex in this example.

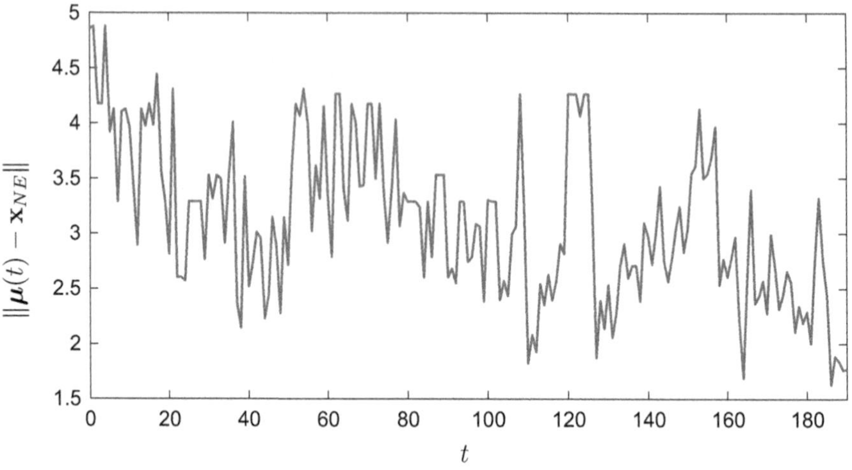

Fig. 4.21 $\|\mu(t) - \mathbf{x}_{\text{NE}}\|$ during the payoff-based algorithm, $N = 100$, $\gamma(t) = \frac{1}{t^{0.6}}$, $\sigma(t) = \frac{1}{t^{0.1}}$

explained by the choice of the rapidly decreasing parameter $\sigma(t)$ that prevents a significant change of the states' values chosen according to the normal distribution with the variance $\sigma(t)$. Figures 4.20 and 4.21 demonstrate behavior of the norm $\|\mu(t) - \mathbf{x}_{\text{NE}}\|$ in the payoff-based learning algorithm with $N = 100$ users, when $\gamma(t) = \frac{1}{t^{0.6}}$ and $\sigma(t) = \frac{1}{t^{0.1}}$. Since the variance is not so small as before, the algorithm needs more time to get close to the Nash equilibrium. However, we do not observe such a slow change in the vector $\mu(t)$ after the first iteration as in the previous simulation. Hence, the tradeoff between the settings for the algorithm's parameter and the convergence rate of the procedure should be analyzed in the future.

4.7 Conclusion

In this chapter the stochastic approximation Robbins–Monro procedure is applied to distributed optimization in multi-agent systems as well as to communication- and payoff-based learning in potential games.

It is started by considering the push-sum protocol that can be adapted to the distributed optimization of the sum of non-convex functions, which are considered local agents' cost functions. The main advantage of this protocol is the minor requirement, namely S-strongly connectivity, on the communication topology. Moreover, to organize communication according to this protocol, agents only need to know the number of the out-neighbors. It is shown that under the push-sum algorithm the agents' local variables converge almost surely to a local solution of the optimization problem (Theorem 4.3.4). For the proof of convergence, the results

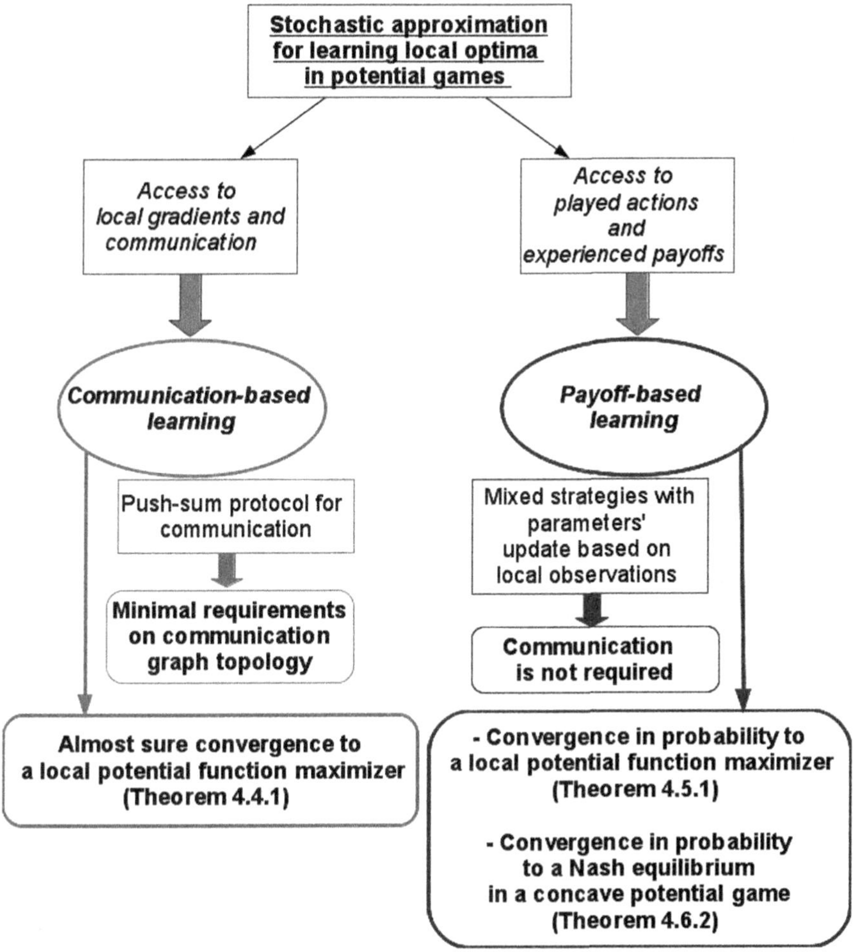

Fig. 4.22 Summary of the results

of the stochastic approximation Robbins–Monro procedure presented in [NK73] are extensively used.

Next, the Robbins–Monro procedure is adapted to learning local maxima of potential functions in continuous action potential games. The obtained results are summarized in Fig. 4.22. Two learning procedures are proposed. The first one is communication-based. Besides communication agents need access to the partial derivative of their utility functions in respect to their own actions. The push-sum protocol is used to set up the exchange of the information between agents. It allows us not to make a restrictive assumption about the communication topology. It is shown that under some rather standard assumptions on the potential function, the joint actions converge almost surely to a local maximum of the potential function (Theorem 4.4.1).

The second algorithm is payoff-based. In this case there is no communication topology in systems. Agents can just observe their individual actual actions and corresponding payoffs. Based on this information, agents choose their actions according to mixed strategies modeled by normal distributions, which parameters are updated during the algorithm's run. Weak convergence and convergence in probability take place in this case (Theorem 4.5.1). Moreover, the payoff-based learning procedure is adapted for learning Nash equilibria maximizing potential functions in concave potential games with a guarantee of an efficient performance (Theorem 4.6.2).

References

[ABSA01] T. Alpcan, T. Basar, R. Srikant, E. Altman, CDMA uplink power control as a noncooperative game, in *Proceedings of the 40th IEEE Conference on Decision and Control, 2001*, vol. 1 (2001), pp. 197–202

[BJ13] P. Bianchi, J. Jakubowicz, Convergence of a multi-agent projected stochastic gradient algorithm for non-convex optimization. IEEE Trans. Autom. Control **58**(2), 391–405 (2013)

[BM06] H. Beigy, M.R. Meybodi, A new continuous action-set learning automaton for function optimization. J. Franklin Inst. **343**(1), 27–47 (2006)

[CLRJ13] A.C. Chapman, D.S. Leslie, A. Rogers, N.R. Jennings, Convergent learning algorithms for unknown reward games. SIAM J. Control Optim. **51**(4), 3154–3180 (2013)

[GHF12] T. Goto, T. Hatanaka, M. Fujita, Payoff-based inhomogeneous partially irrational play for potential game theoretic cooperative control: convergence analysis, in *American Control Conference (ACC), 2012*, June 2012, pp. 2380–2387

[Jen10] M.K. Jensen, Aggregative games and best-reply potentials. Econ. Theory **43**(1), 45–66 (2010)

[KDG03] D. Kempe, A. Dobra, J. Gehrke, Gossip-based computation of aggregate information, in *Proceedings. 44th Annual IEEE Symposium on Foundations of Computer Science, 2003*, October 2003, pp. 482–491

[KNS12] J. Koshal, A. Nedić, U.V. Shanbhag, A gossip algorithm for aggregative games on graphs, in *2012 51st IEEE Conference on Decision and Control (CDC)*, December 2012, pp. 4840–4845

[LM13] N. Li, J.R. Marden, Designing games for distributed optimization. IEEE J. Sel. Top. Sign. Process. **7**(2), 230–242 (2013). Special issue on adaptation and learning over complex networks

[Mal52] I.G. Malkin, *Theory of Stability of Motion*. AEC-tr-3352 Physics and mathematics. Translation series. United States Atomic Energy Commission, Office of Technical Information (1952)

[MB13] I. Matei, J. Baras, A non-heuristic distributed algorithm for non-convex constrained optimization. Institute for Systems Research Technical Reports (2013)

[MS12] J.R. Marden, J.S. Shamma, Revisiting log-linear learning: asynchrony, completeness and payoff-based implementation. Games Econ. Behav. **75**(2), 788–808 (2012)

[NK73] M.B. Nevelson, R.Z. Khasminskii, *Stochastic Approximation and Recursive Estimation*. Translated from the Russian by Israel Program for Scientific Translations ; translation edited by B. Silver (American Mathematical Society, Providence, RI, 1973)

[NO09] A. Nedić, A.E. Ozdaglar, Distributed subgradient methods for multi-agent optimization. IEEE Trans. Autom. Control **54**(1), 48–61 (2009)

[NO14] A. Nedić, A. Olshevsky, Stochastic gradient-push for strongly convex functions on time-varying directed graphs. eprint arXiv:1406.2075 (2014)

[NO15] A. Nedić, A. Olshevsky, Distributed optimization over time-varying directed graphs. IEEE Trans. Autom. Control **60**(3), 601–615 (2015)

[PE09] D.P. Palomar, Y.C. Eldar (eds.), *Convex Optimization in Signal Processing and Communications* (Cambridge University Press, Cambridge, 2009). Cambridge Books Online

[PKL16] D. Paccagnan, M. Kamgarpour, J. Lygeros, On aggregative and mean field games with applications to electricity markets, in *European Control Conference (ECC)*, June 2016

[Pol87] B.T. Poljak, *Introduction to Optimization* (Optimization Software, New York, 1987)

[RM51] H. Robbins, S. Monro, A stochastic approximation method. Ann. Math. Stat. **22**(3), 400–407 (1951)

[RNV10] S.S. Ram, A. Nedić, V.V. Veeravalli, Distributed stochastic subgradient projection algorithms for convex optimization. J. Optim. Theory Appl. **147**(3), 516–545 (2010)

[SBP06] G. Scutari, S. Barbarossa, D.P. Palomar, Potential games: a framework for vector power control problems with coupled constraints, in *2006 IEEE International Conference on Acoustics Speech and Signal Processing Proceedings*, vol. 4, May 2006, pp. 241–244

[SFLS14] G. Scutari, F. Facchinei, L. Lampariello, P. Song, Distributed methods for constrained nonconvex multi-agent optimization-part I: theory. CoRR, abs/1410.4754 (2014)

[SKX15] B. Swenson, S. Kar, J. Xavier, Empirical centroid fictitious play: an approach for distributed learning in multi-agent games. IEEE Trans. Signal Process. **63**(15), 3888–3901 (2015)

[SP14] F. Salehisadaghiani, L. Pavel, Nash equilibrium seeking by a gossip-based algorithm, in *53rd IEEE Conference on Decision and Control*, Dec 2014, pp. 1155–1160

[Tat16a] T. Tatarenko, Stochastic payoff-based learning in multi-agent systems modeled by means of potential games, in *2016 IEEE 55th Conference on Decision and Control (CDC)*, Dec 2016, pp. 5298–5303

[TBA+86] J.N. Tsitsiklis, D.P. Bertsekas, M. Athans et al., Distributed asynchronous deterministic and stochastic gradient optimization algorithms. IEEE Trans. Autom. Control **31**(9), 803–812 (1986)

[TGL13] G. Tychogiorgos, A. Gkelias, K.K. Leung, A non-convex distributed optimization framework and its application to wireless ad-hoc networks. IEEE Trans. Wirel. Commun. **12**(9), 4286–4296 (2013)

[TGM14] T. Tatarenko, L. Garcia-Moreno, A game theoretic and control theoretic approach to incentive-based demand management in smart grids, in *22nd Mediterranean Conference of Control and Automation (MED), 2014*, June 2014, pp. 634–639

[TLR12a] K.I. Tsianos, S. Lawlor, M.G. Rabbat, Consensus-based distributed optimization: practical issues and applications in large-scale machine learning, in *2012 IEEE 50th Annual Allerton Conference on Communication, Control, and Computing (Allerton)*, Oct 2012, pp. 1543–1550

[TS03] A.L. Thathachar, P.S. Sastry, *Networks of Learning Automata: Techniques for Online Stochastic Optimization* (Springer US, New York, 2003)

[TT15] T. Tatarenko, B. Touri, Non-convex distributed optimization. IEEE Trans. Autom. Control **62**(8), 3744–3757 (2017)

[TT16] T. Tatarenko, B. Touri, On local analysis of distributed optimization, in *2016 American Control Conference (ACC)*, July 2016, pp. 5626–5631

[ZC04] V.A. Zorich, R. Cooke, *Mathematical Analysis I*. Mathematical Analysis (Springer, Berlin, 2004)

[ZM13a] M. Zhu, S. Martinez, An approximate dual subgradient algorithm for multi-agent non-convex optimization. IEEE Trans. Autom. Control **58**(6), 1534–1539 (2013)

[ZM13b] M. Zhu, S. Martínez, Distributed coverage games for energy-aware mobile sensor networks. SIAM J. Control Optim. **51**(1), 1–27 (2013)

Chapter 5
Conclusion

This book deals with optimization problems and information processing in multi-agent systems through a game-theoretic formulation and distributed learning. The emphasis is on the networked systems, in which a global optimization problem is formulated. Motivated by possibility to model a range of such problems in terms of potential games in such a way that potential function maximizers correspond to system optima, the work analyzes several learning algorithms applicable to systems with no memory and different information structures. The main goal is to demonstrate that convergence to an optimal (locally optimal) state in any memoryless system can be guaranteed given an appropriate algorithm's setting.

In particular, the work analyzes a general memoryless learning process in systems with the oracle-based information, where agents have discrete action sets. It is shown that such process possesses some guaranteed convergence to a Nash equilibrium coinciding with a potential function maximizer in any discrete action potential game. The convergence rate is estimated for this general learning procedure as well. Furthermore, the work studies an example of these general settings, namely the standard log-linear learning and its modification enabling synchronous updates of agents' actions. It is demonstrated how the parameters of these algorithms can be set up to allow the procedures to converge almost surely to a potential function maximizer. Unfortunately, such settings imply a low convergence rate of the learning procedures. That is why other choices of the parameters are justified to accelerate system's approach to an optimum. Furthermore, the algorithms above are extended to the case of games with continuous actions and studied long-term behavior of the corresponding continuous state dynamics. A particular result consists in introduction of a new method based on the Markov chain approximation technique to describe stochastic stable states of the learning procedures in continuous action games.

Before moving to communication- and payoff-based information structures in modeled games, the work addresses the problem of distributed optimization in networked systems. This problem is of independent interest in context of multi-

T. Tatarenko, *Game-Theoretic Learning and Distributed Optimization in Memoryless Multi-Agent Systems*, DOI 10.1007/978-3-319-65479-9_5

agent systems and can be considered a general case of game-theoretic formulations, where the global objective is to minimize overall costs in games. The book focuses on the push-sum protocol applicable to a general time-variant communication topology without any assumption on double stochastic communication matrices. This algorithm is adapted to the case of non-convex functions and the resulting procedure converges almost surely to a local minimum given some standard assumptions on the goal function. Based on the stochastic approximation technique that turned out to be useful in investigation of the convergence properties of the push-sum algorithm, the work develops the iterative memoryless procedures for learning local optima (local maxima of potential functions) in potential games, where agents can either communicate with each other according to the push-sum protocol or observe only their currently played actions and local actual payoffs. Whereas communication-based approach consists just in a modification of the gradient step in the initial push-sum procedure to steer the average dynamics toward a local maximum of the potential function, the payoff-based approach deals with the agents' mixed strategies, whose parameters are updated by the learning process. As a result, the communication-based learning is shown to converge to a local optimum almost surely. The payoff-based algorithm in its turn is proven to converge weakly to a local maximum of the potential function.

Thus, this book provides a comprehensive analysis of memoryless, and, hence, requiring less technical resources, algorithms which enable solving a wide variety of optimization problems in multi-agent systems with different information structures. Note that all algorithms under consideration are stochastic in their nature and this work successfully applies the theory of Markov processes to study their long-term behaviors.

Appendix A

A.1 Convergence of Probability Measures

This appendix presents a very short overview of the asymptotic behavior of probability measures (distributions). More details on this topic can be found in [Bil99].

We consider here the whole set of the probability distributions Π_d defined on $(\Omega, \sigma(\Omega))$, where Ω is some definite sample space and $\sigma(\Omega)$ is the corresponding Borel sigma-algebra.

Firstly, we introduce the *total variation distance* between two probability distributions.

Definition A.1.1 Let $\mathcal{P}$ and $\mathcal{Q}$ be two distributions from Π_d. Then the *total variation distance* between $\mathcal{P}$ and $\mathcal{Q}$ is defined by

$$\|\mathcal{P} - \mathcal{Q}\|_{\mathrm{TV}} = \sup_{B \in \sigma(\Omega)} |\mathcal{P}(B) - \mathcal{Q}(B)|.$$

We say that the sequence $\{\mathcal{P}_n\}_n \in \Pi_d$ *converges in total variation* to some $\mathcal{P} \in \Pi_d$, if $\|\mathcal{P}_n - \mathcal{P}\|_{\mathrm{TV}} \to 0$ as $n \to \infty$.

The useful theorem for calculation of total variation distances between two absolutely continuous distributions is the following one.

Theorem A.1.1 ([Bil99]) *Let distributions $\mathcal{P}$ and $\mathcal{Q}$ from Π_d have densities $p : \mathbb{R}^d \to \mathbb{R}$ and $q : \mathbb{R}^d \to \mathbb{R}$, respectively (densities with respect to the Lebesgue measure on $\mathbb{R}^d$). Then*

$$\|\mathcal{P} - \mathcal{Q}\|_{\mathrm{TV}} = \frac{1}{2} \int_{\mathbb{R}^d} |p(\mathbf{x}) - q(\mathbf{x})| d\mathbf{x}.$$

T. Tatarenko, *Game-Theoretic Learning and Distributed Optimization in Memoryless Multi-Agent Systems*, DOI 10.1007/978-3-319-65479-9

An equivalent definition of the total variation distance in general case is given by the next lemma.

Lemma A.1.1 ([Bil99]) *For any* $\mathcal{P}, \mathcal{Q} \in \Pi_d$ *the following holds true:*

$$\sup_{B\in\sigma(\Omega)} |\mathcal{P}(B) - \mathcal{Q}(B)| = \frac{1}{2} \sup_{|f|\leq 1} \left| \int_{\Omega} f d\mathcal{P} - \int_{\Omega} f d\mathcal{Q} \right|,$$

where supremum is taken over all Borel functions $f : (\Omega, \sigma(\Omega)) \to (\mathbb{R}, \sigma(\mathbb{R}))$, *whose absolute values are less than* 1.

Apparently, the metric $\|\cdot\|_{\mathrm{TV}}$ "feels" a difference even between very similar distributions. Indeed, let us consider an example, where $\mathcal{P}_n = \delta_{1/n}$ and $\mathcal{P} = \delta_0$. Then, obviously, $\|\mathcal{P}_n - \mathcal{P}\|_{\mathrm{TV}} = 1$ for any n and, thus, no convergence in total variation takes place, although our intuition tells us that $\delta_{1/n}$ should converge in some sense to δ_0 as n tends to infinity. One encounters this problem, because of the broad class of functions f in the definition presented by the lemma above.

To rectify this issue, we consider here another type of convergence of probability measures, namely *weak convergence*.

Definition A.1.2 We say that the sequence $\{\mathcal{P}_n\}_n \in \Pi_d$ *converges weakly* to some $\mathcal{P} \in \Pi_d$, $\mathcal{P}_n \Rightarrow \mathcal{P}$ as $n \to \infty$, if for any Borel function $f : (\Omega, \sigma(\Omega)) \to (\mathbb{R}, \sigma(\mathbb{R}))$ that is continuous and bounded on Ω

$$\int_{\Omega} f d\mathcal{P}_n - \int_{\Omega} f d\mathcal{P} \to 0$$

as $n \to \infty$.

According to the definition above, we can introduce weak topology on Π_d by the following metric:

Definition A.1.3 Let $\mathcal{P}$ and $\mathcal{Q}$ be two distribution from Π_d. Then the *distance between* $\mathcal{P}$ *and* $\mathcal{Q}$ *in weak topology* is defined by

$$\|\mathcal{P} - \mathcal{Q}\|_w = \sup_f \left| \int_{\Omega} f d\mathcal{P} - \int_{\Omega} f d\mathcal{Q} \right|,$$

where supremum is taken over all Borel functions $f : (\Omega, \sigma(\Omega)) \to (\mathbb{R}, \sigma(\mathbb{R}))$, that are continuous and bounded on Ω.

Taking Definitions A.1.1 and A.1.3 into account, we get

$$\|\mathcal{P} - \mathcal{Q}\|_w \leq 2\|\mathcal{P} - \mathcal{Q}\|_{\mathrm{TV}}. \tag{A.1}$$

Thus, convergence in total variation implies weak convergence. The contrary, however, does not hold, as we have seen in the example above.

Important properties, that can be considered equivalent definitions for weak convergence, are formulated in the so-called Portmanteau Lemma [Kle08]. We formulated here only one property figuring in this lemma.

Lemma A.1.2 *The sequence* $\{\mathcal{P}_n\}_n \in \Pi_d$ *converges weakly to some* $\mathcal{P} \in \Pi_d$, $\mathcal{P}_n \Rightarrow \mathcal{P}$ *as* $n \to \infty$, *if and only if*

$$\limsup_{n\to\infty} \mathcal{P}_n(U) \le \mathcal{P}(U)$$

for all closed sets U *from* $\sigma(\Omega)$.

Notice that in the case of discrete probability measures, convergence in weak topology is equivalent to convergence in total variation and means the coordinate-wise convergence of probability vectors.[1]

Other convergence types we deal here with are *almost sure convergence* and *convergence in probability*.

Definition A.1.4 We say that the sequence $\{\mathcal{P}_n\}_n \in \Pi_d$ *converges almost surely* to some $\mathcal{P} \in \Pi_d$, $\mathcal{P}_n \to \mathcal{P}$ a.s. as $n \to \infty$, if

$$\Pr\{\omega \in \Omega : \lim_{n\to\infty} \mathcal{P}_n(\omega) \neq \mathcal{P}(\omega)\} = 0,$$

where Ω is the sample space on which $\{\mathcal{P}_n\}_n$ and $\mathcal{P}$ are defined (in the specific case under consideration $\Omega = \Omega$).

Definition A.1.5 We say that the sequence $\{\mathcal{P}_n\}_n \in \Pi_d$ *converges in probability* to some $\mathcal{P} \in \Pi_d$, $\mathcal{P}_n \to_P \mathcal{P}$ as $n \to \infty$, if for any $\varepsilon > 0$

$$\lim_{n\to\infty} \Pr\{\omega \in \Omega : |\mathcal{P}_n(\omega) - \mathcal{P}(\omega)| > \varepsilon\} = 0,$$

where Ω is the sample space on which $\{\mathcal{P}_n\}_n$ and $\mathcal{P}$ are defined (in the specific case under consideration $\Omega = \Omega$).

Obviously, the almost sure convergence is stronger than convergence in probability (the almost sure convergence implies convergence in probability [Shi95]). Convergence in probability, in its turn, implies weak convergence. The contrary is, however, not true. Nevertheless, if probability measures defined on $\Omega = \Omega$ converge weakly to a singular measure $\delta_{\boldsymbol{a}}$, $\boldsymbol{a} \in \Omega$, then this convergence is also in probability [Shi95]:

Theorem A.1.2 *Let* $\{\mathcal{P}_n\}_n$ *be defined on* $(\Omega, \sigma(\Omega))$. *If* $\mathcal{P}_n \Rightarrow \delta_{\boldsymbol{a}}$, *then* $\mathcal{P}_n \to_P \delta_{\boldsymbol{a}}$.

[1] Probability (stochastic) vector is any vector with nonnegative coordinates whose sum is 1.

A.2 Markov Chains with Finite States

This appendix contains the main notations and results regarding the theory of Markov chains with discrete states. These results are used to prove the main results in Chap. 3, Sects. 3.2–3.5.

A Markov chain is a memoryless stochastic process $\{\xi_t\}_{t\in\mathbb{Z}^+}$ with discrete time. Let some finite set $S = \{s_1, \dots, s_m\}$ be the state set of this process. Then

$$\Pr\{\xi_t = s_{k_t}|\xi_{t-1} = s_{k_{t-1}}, \xi_{t-2} = s_{k_{t-2}}, \dots, \xi_0 = s_{k_0}\} = \Pr\{\xi_t = s_{k_t}|\xi_{t-1} = s_{k_{t-1}}\}$$

for all possible $s_{k_j} \in S$, $j = 0, \dots, t$. Thus, the transition of the process from one state $s_i \in S$ to another state $s_j \in S$ can be defined by the transition probability

$$p_{ij}(t) = \Pr\{\xi_t = s_j|\xi_{t-1} = s_i\}.$$

Apparently, $\sum_{j=1}^m p_{ij}(t) = 1$ for any $t \in \mathbb{Z}^+$ and any $i \in [m]$. This allows us to associate the matrix $P(t) = (p_{ij}(t))_{i,j\in[m]}$ with the Markov process $\{\xi_t\}_{t\in\mathbb{Z}^+}$. This matrix is called transition probability matrix and defines the probability for the chain to transit from any state $s_i \in S$ to any state $s_j \in S$ at the moment of time t. Let $\boldsymbol{\pi}(0) = (\pi^{s_1}(0), \dots, \pi^{s_m}(0))$, $\sum_{j=1}^m \pi^{s_j}(0) = 1$, be the m-dimensional probability vector corresponding to the initial distribution of the Markov chain (initial state of the Markov chain). In other words, the distribution $\boldsymbol{\pi}(0)$ is the distribution of ξ_0 over S, and each coordinate $\pi^{s_i}(0)$ defines the probability for the chain to be initiated in the state s_i. Then, according to the law of total probability, the distribution of the chain at the time $t = 1$ is

$$\boldsymbol{\pi}(1) = \boldsymbol{\pi}(0)P(0).$$

Going further, we get the state of the Markov chain after t steps:

$$\boldsymbol{\pi}(t) = \boldsymbol{\pi}(t-1)P(t-1) = \boldsymbol{\pi}(0)\prod_{k=0}^{t-1} P(k),$$

where $\boldsymbol{\pi}(t)$ is the distribution of the Markov chain at time t.

Let $P_{t,n} = \prod_{k=t}^{t+n-1} P(k) = (p_{ij}(t,n))_{i,j\in[m]}$. Thus, each element $p_{ij}(t,n)$ of the matrix $P_{t,n}$ defines the probability for the Markov chain, being at the moment t in the state s_i, to transit to the state s^j after n steps.

A.2.1 Time-Homogeneous Markov Chains with Finite States

In the case of time-homogeneous Markov chains the transition probability matrix is a constant matrix, i.e., for any t

$$P(t) = P \text{ and } P_{t,n} = P^n,$$

where n is the power. Thus, the probabilities $p_{ij}(t, n) = p_{ij}(n)$, $i, j \in [m]$, depend only on n.

The state $s_j \in S$ is said to be accessible from the state $s_i \in S$ ($s_i \to s_j$), if there exists an integer $n_{ij} \geq 1$ such that $p_{ij}(n_{ij}) > 0$. Every state is defined to be accessible from itself. The state s_i is said to *communicate* with the state s_j, if both $s_i \to s_j$ and $s_j \to s_i$. The set of states C is a *communicating class* if every pair of states in C communicates with each other, and no state in C communicates with any state not in C. A Markov chain is said to be *irreducible*, if its state space is a single communicating class. The state s_i has the period k_i, if $k_i = \gcd\{n : p_{ii}(n) > 0\}$.[2] If $k_i = 1$, the state i is called *aperiodic*.

Definition A.2.1 A time-homogeneous irreducible Markov chain with aperiodic states is called *ergodic*.

Another equivalent definition of the ergodicity is the following.

Definition A.2.2 A time-homogeneous Markov chain is *ergodic*, if there exists such $n_0 > 0$ that $\inf_{i,j} p_{ij}(n_0) > 0$.

Now we define stationary distributions of a Markov chain.

Definition A.2.3 The probability vector $\boldsymbol{\pi}$ is called *stationary distribution* of the Markov chain P if $\boldsymbol{\pi} P = \boldsymbol{\pi}$.

Note that any Markov chain has a stationary distribution, but this distribution might not be unique. However, in the case of an irreducible Markov chain this distribution is unique. Moreover, for any ergodic Markov chain the following theorem is fulfilled [Fel68]:

Theorem A.2.1 *Any ergodic Markov chain P on the state space S has the unique stationary distribution $\boldsymbol{\pi}$. Moreover, the coordinate π^{s_j} of this vector is the limit of $p_{ij}(n)$ as $n \to \infty$, i.e., $\lim_{n\to\infty} p_{ij}(n) = \pi^{s_j}$ for all i.*

Thus, any ergodic Markov chain has the property to possess the unique "final" distribution vector $\boldsymbol{\pi}$, which is the stationary distribution. In other words,

$$\lim_{n\to\infty} \boldsymbol{\pi}(n) = \lim_{n\to\infty} \boldsymbol{\pi}(0)P^n = \boldsymbol{\pi}$$

for any initial distribution vector $\boldsymbol{\pi}(0)$. The matrix of transition probabilities in any ergodic homogeneous Markov chain is called *regular*.

A useful expression for the coordinates of the unique stationary distribution of an ergodic Markov chain P can be found in [FW84] (Lemma 3.1 of Chapter 6). We denote by $G(s)$ the set of s-graphs,[3] the letter g denotes a graph. For any g the

[2] "gcd" is the greatest common divisor.

[3] A graph consisting of arrows $s' \to s''$, $s' \in S \setminus s$, $s'' \in S$, $s' \neq s''$ is s-graph, if there are no closed cycles in the graph and every point $s' \in S \setminus s$ is the initial point of exactly one arrow.

expression $\Pi(g)$ denotes $\prod_{(s_i \to s_j) \in g} p_{ij}$ with $p_{ij} : i, j \in S, j \neq i$. Then the following holds true.[4]

Theorem A.2.2 *Let us consider an ergodic Markov chain with a set of states $S = (s_1, \ldots, s_m)$ and a transition probability matrix P. Then the unique stationary distribution π has the following coordinates:*

$$\pi^{s_i} = \frac{Q_i}{\sum_{s_j \in S} Q_j}, \quad s_i \in S,$$

where

$$Q_k = \sum_{g \in G(s_k)} \Pi(g), \quad k = 1, \ldots, s_m$$

Thus, any ergodic time-homogeneous Markov chain $\{\xi_t\}_t$ defined on the finite set S by some regular matrix P converges in total variation to its unique stationary distribution π. It implies

$$\lim_{t \to \infty} \Pr\{\xi_t = s_i\} = \pi^{s_i}, \quad s_i \in S,$$

where π^{s_i} is defined in the theorem above.

A.2.2 Time-Inhomogeneous Markov Chains with Finite States

For a time-inhomogeneous Markov chain there exist two notions of ergodicity: weak and strong ergodicity [HB58]. In this subsection we deal with time-inhomogeneous chains on a finite state space S.

Definition A.2.4 A Markov chain $\{P(t)\}$ is *weakly ergodic*, if for any t and all i, i', j, $|p_{ij}(t,n) - p_{i'j}(t,n)| \to 0$ as $n \to \infty$. Moreover, if the elements of $P_{t,n}$ tend to some limits as $n \to \infty$ and these limits are the same for all t, then the Markov chain $P(t)$ is *strongly ergodic*.

In the case of some strongly ergodic Markov chain there exists a unique probability row-vector π^* such that for any initial distribution vector π_0 on S and for any t we have $\lim_{n \to \infty} \pi_0 P_{t,n} = \pi^*$. We call this vector *stationary distribution* of the strongly ergodic inhomogeneous Markov chain under consideration. Note that weak and strong ergodicity are equivalent in the case of time-homogeneous Markov chains.

Another notion that we use further is the *scrambling matrix* [HB58].

[4]Note that the result holds not only for an ergodic Markov chain, but also for any irreducible one (see [FW84], Lemma 3.1 of Chapter 6).

Definition A.2.5 A matrix P is called *scrambling matrix*, if for any two rows, say i and i', there exists at least one column, j, such that both $p_{ij} > 0$ and $p_{i'j} > 0$.

The weak ergodicity of time-inhomogeneous Markov chains can be studied by using a concept of the coefficient of ergodicity introduced by Dobrushin [Dob56].

Definition A.2.6 The coefficient of ergodicity of some transition probability matrix P is denoted by $\tau(P)$ and is defined as $\tau(P) = 1 - \min_{i,i'} \sum_j \min(p_{ij}, p_{i'j})$.

It follows directly from Definition A.2.6 that $1 - \tau(P) = 0$, if and only if P is not a scrambling matrix. *Pattern* of a matrix is an arrangement of non-zero elements in it. If a stochastic matrix P is regular, i.e., corresponds to some ergodic time-homogeneous Markov chain, then there exists an integer $c < \infty$ such that P^c is a scrambling matrix. It is clear that the patterns of matrices completely determine the pattern of their product. That allows us to formulate the next lemma.

Lemma A.2.1 *If the transition matrices $P(t)$ of some inhomogeneous Markov chain are regular and contain the same pattern for every t, then there exists an integer c such that $P_{t,c} = \prod_{i=t}^{t+c-1} P(i)$ is a scrambling matrix for all t, and, hence, $1 - \tau(P_{t,c}) > 0$.*

The minimal constant among such constants c is called *scrambling constant*.

The following two lemmata [Dob56] can be useful in analysis of the coefficient of ergodicity.

Lemma A.2.2 *For a transition probability matrix P defined on $S = \{s_1, \dots, s_m\}$, and any vector $\boldsymbol{\mu} \in \mathbb{R}^m$ such that $\sum_{i=1}^{m} \mu^i = 0$, the following is fulfilled:*

$$\|\boldsymbol{\mu} P\|_{l_1} \leq \tau(P)\|\boldsymbol{\mu}\|_{l_1}.$$

Lemma A.2.3 *Let P and Q be any two transition probability matrices. Then $\tau(PQ) \leq \tau(P)\tau(Q)$.*

Finally, we formulate the following well-known results on a criterion for weak ergodicity and sufficient conditions for strong ergodicity presented in [HB58] and [IM76], respectively.

Theorem A.2.3 *A Markov chain $P(t)$ is weakly ergodic if and only if there exists a subdivision of the sequence of steps into blocks at steps numbered $t_1, t_2, t_3, \dots$, such that*

$$\sum_{j=1}^{\infty} \alpha(P_{t_j,n_j}) = \infty,$$

where $n_j = t_{j+1} - t_j$ and $\alpha(P_{t_j,n_j}) = 1 - \tau(P_{t_j,n_j})$.

Theorem A.2.4 *[IM76] A time-inhomogeneous Markov chain $P(t)$ is strongly ergodic, if the following conditions hold:*

(1) The Markov chain $P(t)$ is weakly ergodic,
(2) For each fixed t, the time-homogeneous Markov chain $P(t)$ for any fixed t is ergodic with a stationary distribution $\pi(t)$.
(3) $\sum_{t=0}^{\infty} \|\pi(t) - \pi(t+1)\|_{l_1} < \infty$.

Moreover, if $\pi^* = \lim_{t\to\infty} \pi(t)$, then π^* is the unique stationary distribution of the time-inhomogeneous Markov chain $P(t)$.

A.3 Markov Chains with General States

Now we turn our attention to Markov chains defined on a general metric state space X with the Borel sigma-algebra of all subsets $\sigma(X)$. Any Markov chain (possibly time-inhomogeneous) $\{\xi_t\}_t$ on X is defined by *stochastic transition kernel*:

$$P_t : X \times \sigma(X) \to [0, 1].$$

The transition kernel $P_t(x, B) = P^{t,1}(x, B)$ is the probability for the chain being at $x \in X$ at time t to transit to the set $B \in \sigma(X)$. (The kernel $P^{t,1}$ should be regarded as the 1-step transition probability function at the moment t.) Any transition probability kernel acts from the right on any probability measure μ and from left on any bounded measurable function f defined on X, namely

$$\mu P_t(\cdot) = \int_X P_t(x, \cdot)\mu(dx) \text{ and } P_t f(x) = \int_X f(y) P_t(x, dy).$$

The kernel evolution over time k from the moment t is defined by the following equality:

$$P^{t,k}(x, B) = \int_X P^{t,m}(x, dy) P^{t+m,k}(y, B), \quad 1 \le m \le k-1.$$

$P^{0,k}$ is further denoted by P^k.

A.3.1 Time-Homogeneous Markov Chains with General States

Let us now focus on time-homogeneous Markov chains, i.e., $P_t(x, \cdot) = P(x, \cdot)$ and, hence, $P^{t,k} = P^k$ does not depend on t. We say that some measure μ defined on the measurable space $(X, \sigma(X))$ is *stationary* for the Markov chain $P(x, \cdot)$, if $\mu P = \mu$.

Definition A.3.1 A Markov chain $P(x,\cdot)$ is *ϕ-irreducible*, if there exists a non-zero σ-finite measure ϕ on X such that for all $A \in \sigma(X)$ with $\phi(A) > 0$ and for all $x \in X$ there exists a positive integer $k = k(x, A)$ such that $P^k(x, A) > 0$.

Definition A.3.2 A ϕ-irreducible Markov chain $P(x,\cdot)$ is *recurrent*, if for every $x \in X$ and every $A \in \sigma(X)$ with $\phi(A) > 0$, $\sum_{k=1}^{\infty} P^k(x, A) = \infty$.

Roughly speaking, Markov chains falling under Definition A.3.1 have positive probability of eventually reaching any subset with some positive ϕ-measure from any state $x \in X$. The proof of the following theorem can be found in [MT09].

Theorem A.3.1 *Any ϕ-irreducible recurrent Markov chain has a* unique *(up to a constant multiplier) stationary measure.*

If such measure is finite, then the chain is called *ergodic*. The following sufficient condition for ergodicity was formulated in [Doo53].

Theorem A.3.2 *A Markov chain $P(x,\cdot)$ defined on a separable metric space X is ergodic, if there exists a probability measure θ on $\sigma(X)$, a fixed integer $k > 0$, and $\delta > 0$, such that $P^k(x, A) \le 1 - \delta$, whenever $\theta(A) \le \delta$ for $A \in \sigma(X)$.*

Another way to check the existence of the stationary distribution is to prove the *reversibility* of the Markov chain.

Definition A.3.3 A Markov chain $P(x,\cdot)$ defined on a separable metric space X is *reversible* with respect to a probability distribution Π defined on $(X, \sigma(X))$, if

$$\Pi(dx)P(x, dy) = \Pi(dy)P(y, dx). \tag{A.2}$$

Equation (A.2) is also called *detailed balance equation*. A very important property of reversibility is the following one.

Theorem A.3.3 *If Markov chain is reversible with respect to Π, then Π is a stationary measure for the chain.*

However, convergence over time to the stationary distribution may fail due to the chain's periodicity. Moreover, a slightly stronger recurrence condition, namely Harris recurrence, is required [MT09].

Definition A.3.4 A Markov chain $P(x,\cdot)$ with some stationary distribution Π is *aperiodic*, if there do not exist $d \ge 2$ and disjoint subsets $X_1, X_2, \ldots, X_d \in \sigma(X)$ with $P(x, X_{i+1}) = 1$ for all $x \in X_i$ $(1 \le i \le d-1)$, and $P(x, X_1) = 1$ for all $x \in X_d$, such that $\Pi(X_1) > 0$. Otherwise, the chain is *periodic* with period d and periodic decomposition $X_1, X_2, \ldots, X_d$.

Let $\xi(n)$ be a random variable that corresponds to the state of the Markov chain after n steps.

Definition A.3.5 A Markov chain $P(x,\cdot)$ with some stationary distribution Π is *Harris recurrent*, if for all $x \in X$ and all $A \in \sigma(X)$ with $\Pi(A) = 0$,

$$\Pr\{\xi(n) \in A \text{ for all } n | \xi(0) = x\} = 0.$$

Now we formulate the theorem that claims that a ϕ-irreducible, aperiodic, Harris recurrent Markov chain has a unique stationary measure and P^k converges to this measure in total variation as $k \to \infty$ [MT09].

Theorem A.3.4 *If a Markov chain $P(x,\cdot)$ on a separable metric space X is ϕ-irreducible, aperiodic, Harris recurrent, then it has the unique stationary measure Π and*

$$\lim_{k\to\infty} \|P^k(x,\cdot) - \Pi(\cdot)\|_{\mathrm{TV}} = 0$$

for all $x \in X$.

A.3.2 Convergence of Transition Probability Kernels

In this subsection we discuss some convergence properties of transition probability kernels and present an important result obtained in [Pin92]. As before, we deal with a metric separable space X. Let us consider a sequence of transition probability kernels $\{P_n\}$ and a kernel P on X. Recall that $\Rightarrow$ denotes weak convergence of measures (see Appendix A.1).

Definition A.3.6 We say that P_n converges to P, $P_n \to P$ as $n \to \infty$, if for any measures $\{\mu_n\}$, μ such that $\mu_n \Rightarrow \mu$ on the space $(X, \sigma(X))$ as $n \to \infty$, the following is fulfilled:

$$\mu_n P_n \Rightarrow \mu P \text{ as } n \to \infty.$$

The following lemma (see [Pin92] for the proof) formulates two conditions that are together sufficient for transition kernel convergence in sense of Definition A.3.6.

Lemma A.3.1 *Assume that $\{P_n\}$ and P satisfy the following two conditions:*

(1) $g(x) = \int_X f(y)P(x,dy)$ is a continuous function of $x \in X$ for every continuous bounded function f on X.
(2) for any $x \in X$ $P_n(x,\cdot) \Rightarrow P(x,\cdot)$ as $n \to \infty$.

Then $P_n \to P$ as $n \to \infty$.

Now we formulate the theorem (see [Pin92] for the proof) eliminating the relation between stationary measures of convergent stochastic kernels.

Theorem A.3.5 *Let X be a compact subset of $\mathbb{R}^d$, transition probability kernels $\{P_n\}$, P be defined on $(X, \sigma(X))$, and $P_n \to P$ as $n \to \infty$. Let the kernel P have a unique stationary measure μ and the kernel P_n have a stationary measure μ_n, which is not necessarily unique. Then $\mu_n \Rightarrow \mu$ as $n \to \infty$.*

A.3.3 Time-Inhomogeneous Markov Chains with General States

We consider a Markov chain on some separable space $(X, \sigma(X))$. Let this chain be defined by the transition probability (stochastic) kernel $P_t(x, \cdot)$ that is time dependent. We remind here that $P_t(x, \cdot)$ is a measure on $(X, \sigma(X))$ for any fixed $x \in X$, such that $P_t(x, B)$ is a Borel-measurable function of x for any $B \in \sigma(X)$. Thus, $P_t(x, B) = P^{t,1}(x, B)$ is the probability for the chain to be in the set B at the step $t + 1$ being in the state x at the tth step.

Let $\Lambda(X)$ denote the set of all probability measures defined on $(X, \sigma(X))$. The convergence properties of the most general inhomogeneous Markov chains can be studied by using a concept of the coefficient of ergodicity introduced by Dobrushin [Dob56].

Definition A.3.7 The real number

$$\tau(P) = \sup_{\lambda,\mu \in \Lambda(X), \lambda \neq \mu} \frac{\|\mu P - \lambda P\|_{\mathrm{TV}}}{\|\mu - \lambda\|_{\mathrm{TV}}}$$

is called the *coefficient of ergodicity* of the transition probability kernel P.

In the seminal work [Dob56] other formulas which are equivalent to one in the definition above are presented. For example,

$$\tau(P) = 1 - \inf_{x', x''} \inf_{\{X_i\}} \sum_{i=1}^{l} \min(P(x', X_i), P(x'', X_i)), \tag{A.3}$$

where the first infimum is taken over $x', x'' \in X$, the second infimum is taken over all possible partitions of X into pairs of the non-intersecting and non-empty subsets $X_i, i = 1, \ldots, l$, and the number of subsets $l \geq 2$ can be chosen arbitrarily. Another formula is

$$\tau(P) = \sup |P(x', B) - P(x'', B)|, \tag{A.4}$$

where the supremum is taken over all $x', x'' \in X$ and all $B \in \sigma(X)$.

Analogously to the case of finite state Markov chains, the following three lemmata regarding the coefficient of ergodicity can be formulated [MI73].

Lemma A.3.2 *For a kernel function $R(x, \cdot)$ defined on $X \times \sigma(X)$ and such that $\int_X R(x, dy) = 0$, and any stochastic kernel P the following is fulfilled:*

$$\|PR\|_{\mathrm{TV}} \leq \tau(P) \|R\|_{\mathrm{TV}}.$$

Lemma A.3.3 *Let P and Q be any two stochastic kernels. Then $\tau(PQ) \leq \tau(P)\tau(Q)$.*

Lemma A.3.4 *If the transition kernel P_t of some inhomogeneous Markov chain is such that for every fixed k the time-homogeneous chain P_k satisfies all conditions in Theorem A.3.4, then there exists an integer c such that $1 - \tau(P^{k,c}) > 0$ for all k.*

Now we formulate the concepts of weak and strong ergodicity for time-inhomogeneous Markov chains with general states.

Definition A.3.8 The Markov chain P_t defined on X is *weakly ergodic*, if for any $m \in \mathbb{N}$

$$\lim_{k\to\infty} \sup_{\lambda,\mu\in\Lambda(X)} \|\mu P^{m,k} - \lambda P^{m,k}\|_{\mathrm{TV}} = 0.$$

Analogously to the discrete state Markov chains, weak ergodicity means that the information on the initial state as well as the information on any distant past vanishes with time. But it does not imply convergence of $\lambda P^{m,k}$ for any $\lambda \in \Lambda(X)$ to some measure from $\Lambda(X)$ in total variation. If such convergence takes place, the Markov chain is called *strongly ergodic*.

Definition A.3.9 The Markov chain P_t defined on X is *strongly ergodic*, if there exists a distribution $\pi \in \Lambda(X)$, called limit distribution, such that for any $m \in \mathbb{N}$

$$\lim_{k\to\infty} \sup_{\lambda\in\Lambda(X)} \|\lambda P^{m,k} - \pi\|_{\mathrm{TV}} = 0.$$

The criterion of weak ergodicity analogous to one in Theorem A.2.3 can be formulated for Markov chains with general states [MI73].

Theorem A.3.6 *The Markov chain P_t defined on X is weakly ergodic, if and only if there exists a subdivision of the sequence of steps into blocks at steps numbered t_1, t_2, $t_3, \ldots,$ such that*

$$\sum_{j=1}^{\infty} \alpha(P^{t_j,n_j}) = \infty,$$

where $n_j = t_{j+1} - t_j$ and $\alpha(P^{t_j,n_j}) = 1 - \tau(P^{t_j,n_j})$.

To prove the strong ergodicity of some general state Markov chain, one can use the following result [IM76].

Theorem A.3.7 *The Markov chain P_t defined on X is strongly ergodic, if the following conditions hold:*

(1) P_t is weakly ergodic,
(2) There is an equilibrium distribution of P_k for each $k \in \mathbb{N}$, i.e., there exists such $\pi_k \in \Lambda(X)$ that $\pi_k P_k = \pi_k$,
(3) $\sum_{k=1}^{\infty} \|\pi_k - \pi_{k+1}\|_{\mathrm{TV}} < \infty$.

Moreover, if $\pi^* \in \Lambda(X)$ is such that $\lim_{k\to\infty} \|\pi_k - \pi^*\|_{\mathrm{TV}} = 0$, then π^* is the limit distribution of the time-inhomogeneous Markov chain P_t.

References

[Bil99] P. Billingsley, *Convergence of Probability Measures*. Wiley Series in Probability and Statistics: Probability and Statistics, 2nd edn. (Wiley, New York, 1999). A Wiley-Interscience Publication

[Dob56] R.L. Dobrushin, Central limit theorem for non-stationary Markov chains I, II. Theory Probab. Appl. **1**, 65–80 (1956)

[Doo53] J.L. Doob, *Stochastic Processes* (Wiley, New York, NY, 1953)

[Fel68] W. Feller, *An Introduction to Probability Theory and Its Applications*. Wiley Series in Probability and Mathematical Statistics (Wiley, New York, 1968)

[FW84] M.I. Freidlin, A.D. Wentzell, *Random Perturbations of Dynamical Systems*. Grundlehren der mathematischen Wissenschaften (Springer, Berlin, 1984)

[HB58] J. Hajnal, M.S. Bartlett, Weak ergodicity in non-homogeneous Markov chains. Math. Proc. Camb. Philos. Soc. **54**, 233–246 (1958)

[IM76] D.L. Isaacson, R.W. Madsen, *Markov Chains, Theory and Applications*. Wiley Series in Probability and Mathematical Statistics (Wiley, New York, 1976)

[Kle08] A. Klenke, *Probability Theory: A Comprehensive Course* (Springer, London, 2008)

[MI73] R.W. Madsen, D.L. Isaacson, Strongly ergodic behavior for non-stationary Markov processes. Ann. Probab. **1**(2), 329–335 (1973)

[MT09] S. Meyn, R.L. Tweedie, *Markov Chains and Stochastic Stability*, 2nd edn. (Cambridge University Press, New York, NY, 2009)

[Pin92] S.M. Pincus, Approximating Markov chains. Proc. Natl. Acad. Sci. **89**(10), 4432–4436 (1992)

[Shi95] A.N. Shiryaev, *Probability*, 2nd edn. (Springer, New York, Secaucus, NJ, 1995)

Printed in the United States
By Bookmasters